AN EYE
ON FLANDERS

AN EYE ON FLANDERS: THE GRAPHIC ART OF JULES DE BRUYCKER

by Stephen H. Goddard

The University of Kansas
Lawrence, Kansas 66045

AN EYE ON FLANDERS: THE GRAPHIC ART OF JULES DE BRUCKYER (1870–1945)

Exhibition organized by Stephen H. Goddard.
Publication of this exhibition catalogue is funded in part by a contribution from The Breidenthal-Snyder Foundation.

Exhibition itinerary

Spencer Museum of Art
University of Kansas, Lawrence
March 30–May 19, 1996

Elvehjem Museum of Art
University of Wisconsin-Madison
September 7–November 10, 1996

Georgia Museum of Art
University of Georgia, Athens
December 14, 1996–January 25, 1997

Managing editor: Sally Hayden
Graphic designer: Valerie Spicher
Photographer: Robert Hickerson

© 1996 by the Spencer Museum of Art, University of Kansas, Lawrence.
All rights reserved. No part of this work may be reproduced or transmitted in any form or by any means, electronic or mechanical, including photocopying and recording, or by any information storage or retrieval system without permission in writing from the publisher.

First edition, published 1996.
Printed in the United States of America.
10 9 8 7 6 5 4 3 2 1
ISBN 0-913689-39-4

For John Talleur

TABLE OF CONTENTS

Foreword	vi
Acknowledgments	vii
Preface	viii
List of Lenders	x
Map of Ghent	xi
An Eye on Flanders: The Graphic Art of Jules De Bruycker	1
The City Context	3
Class Consciousness and Self-Consciousness	7
Patershol	9
The Early Etchings	15
The Great War	21
Recognition and the Second World War	31
Catalogue	49
Bibliography	103
Index	109

FOREWORD

Jules De Bruycker's work is dense, amusing, frightening, satirical, whimsical, and powerful. He depicted with biting accuracy the drama and social life of Flanders during the first half of the twentieth century; his career spans the two World Wars. Though De Bruycker is not widely known in the United States, he has been an interest of Spencer Museum Curator Stephen Goddard for several years. In his hometown of Ghent, De Bruckyer was always referred to as a *tapissier* or *behangersgast* (upholsterer or paper hanger), even when he was a successful artist elsewhere. To his colleagues he may have seemed an ordinary bourgeois, but his work displays strong visual powers and extraordinary understanding and criticism of the culture in which he lived.

Goddard's essay places De Bruycker in both the Flemish community in Belgium and the artistic community in Ghent, so that this catalogue and exhibition provide a view on early twentieth-century Belgium life. De Bruycker depicted his surroundings in the medium of the graphic arts, particularly etching. The brutal humor of some of his imagery contrasts to the more romanticized view of poverty characteristic of much of turn-of-the-century realism.

Andrea S. Norris
Director, Spencer Museum of Art

ACKNOWLEDGMENTS

This project would not have been possible without the research supported by grants from the Franklin D. Murphy Travel Fund at the University of Kansas. A number of colleagues in Belgium and the United States have openly shared their knowledge and I thank them: John De Bruycker; Brendan Burny; Roger Cardon; Marjorie Cohn of the Fogg Museum of Art; Micheline Colin of the Archives de l'art contemporain, Brussels; John Convent; Jan Dewilde of the Stedelijke Musea Ieper; Luc and Adrienne Fontainas; Robert Hoozee and Moniek Nagels of the Museum voor Schone Kunsten Gent; Ben Johnson of the British Museum, London; Wouter Steenhaut of the Archief en Museum van de Socialistische Arbeidersbeweging, Ghent; Mme. Van de Perre; and Nicole Walch of the Bibliothèque royale Albert 1er, Brussels. Special thanks also to Willy Juwet, R. Elson, and Bruno Hicguet of the Ministerie van de Vlaamse Gemeenschap, Department Welzijn, Volksgezondheid en Cultuur (Ministry of the Flemish Community, Department of Public Health and Culture) who generously sponsored the international shipment of the works of art.

At the Spencer Museum of Art I especially thank Director Andrea Norris, for her willingness to support an exhibition of graphic art by a virtually unknown artist, and my interns, Bill North, Jill Vessely, Stacey Skold, and Sarah Crawford, for their help over the past few years. Thanks also to Susan Craig and the staff of the Murphy Library of Art and Architecture.

Warm appreciation goes to friends who have made my visits and research in Belgium go smoothly: John De Bruycker, Lode De Clercq and Nadine Tasell, Robert and Hedwig Hoozee, Milt Papatheofanis, and Maurice Tzwern. Finally, I am grateful to my family, Diane, Erica, Emily, and Caitlin, for their patience during my periods of study abroad.

Stephen H. Goddard
Curator of Prints and Drawings, Spencer Museum of Art

PREFACE

This study began after Robert Hoozee showed me the holdings of prints and drawings by Jules De Bruycker at the Museum voor Schone Kunsten in Ghent. I soon realized that these prints were by the same artist that Richard Field, of the Yale University Art Gallery, once suggested I might find interesting. While De Bruycker has been almost over exposed in Belgium, he is virtually unknown here in the United States, and I became determined to take on the task of an introductory exhibition. At first I planned to find a way to deal with De Bruycker in a group study, but as I became more familiar with the material it became clear that De Bruycker was in many ways a singular phenomenon and I resolved to form a monographic study. The Spencer Museum of Art and the Museum voor Schone Kunsten in Ghent considered collaborating on a De Bruycker exhibition. But as the Ghent exhibition was planned to commemorate the fiftieth anniversary of the artist's death to an audience already intimately familiar with Ghent and its traditions, we resolved to assemble separate exhibitions from a similar group of works, but with substantially different catalogue texts. The choice of works for this exhibition is fairly orthodox in that it sets out to define the major contours of the artist's career as a printmaker and draftsman, not to pause and examine the precious or unusual aspects of his output, or the multiplicity of states and impressions of his more heavily worked plates.

The De Bruycker literature is surprisingly large, but it is not easily accessible. Probably the best study to date on De Bruycker is Georges Chabot's long article, written in French but published in Flemish in the *Kultureel Jaarboek voor de Provincie Oostvlaanderen* [Cultural Annual for the Province of East Flanders]. A study by Achilles Mussche provides a more personal and evocative view, and the catalogue by Paul Eeckhout for the artist's centenary exhibition at the Museum voor Schone Kunsten in Ghent includes important statements by those who knew De Bruycker. The standard catalogue raisonné by Grégoire Le Roy includes a useful study, but as a catalogue it is somewhat incomplete and, of course, totally lacking for works produced after its 1933 publication. By the time this catalogue is in print, several exhibitions commemorating the fiftieth anniversary (1995) of De Bruycker's death will have appeared; both the Museum voor Schone Kunsten Gent and the Stedelijk Museum van Ieper are producing exhibition catalogues.

Since the present study is primarily chronological, it follows the main contours established by Chabot, Eeckhout, and Le Roy. However, since this text is for a non-Belgian audience, we have taken the opportunity to address a number of topics along the way that might be taken for granted in De Bruycker's home territory.

A note on language: De Bruycker gave his prints various titles in both French and Flemish. In the text we refer to the title as it appears on the exhibited impression of the print, but if the exhibited impression is not inscribed with a title, the title is generally given as it appears in Le Roy's catalogue raisonné. Similarly, when reference is made to a print that is not exhibited, the title recorded by Le Roy is used. Flemish place names are given in Flemish (Ieper, rather than Ypres, for example) unless there is a conventional English spelling, such as Ghent, Brussels, or Antwerp. Place names in Ghent, including names of churches, are given in Flemish (Sint-Niklaas, instead of Saint Nicholas, for example). In instances where books have been simultaneously published in French and Flemish, only the edition consulted is given in the bibliography.

All translations are the author's unless otherwise noted; original texts are given in the endnotes. While several colleagues have made suggestions and corrections regarding these translations, any errors are my own.

Stephen H. Goddard

LIST OF LENDERS

Bibliothèque Royale Albert 1er, Cabinet des Estampes, Brussels
J. De Bruycker Collection
The Harvard University Art Museums, Cambridge
Museum voor Schone Kunsten, Ghent
Spencer Museum of Art, The University of Kansas, Lawrence

GHENT

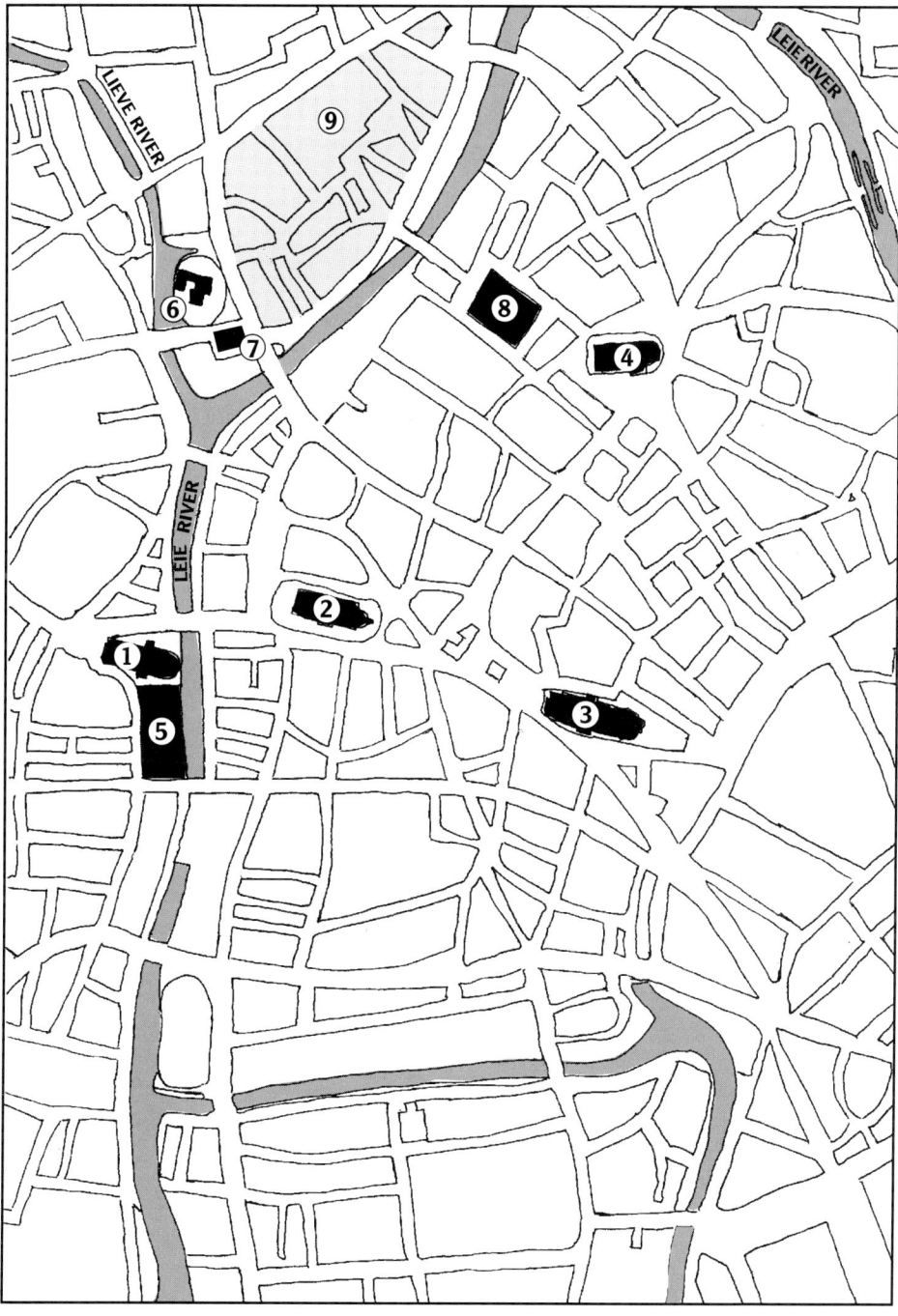

1. Sint-Michiels
2. Sint-Niklaas
3. Sint-Baafs
4. Sint-Jacobs
5. het Pand (Dominican Pand)
6. Gravensteen (Castle of the Counts)
7. Sint-Veerleplein (St. Veerle Square)
8. Vrijdagmarkt (Friday Market)
9. Patershol District
   (with the Carmelite Pand)

"De Bruycker may also be called an eye. But an eye in which the optic nerve plunges into self-denying discontent, into self-mocking bitterness. [Ook De Bruycker mocht worden genoemd : een Oog. Maar een oog, waarvan de zenuwen-bundel duikt in misnoegdheden die zich-zelf verloochenen, in verbittering die zich-zelf bespot.]"

Karel van de Woestijne, 1912

Jules De Bruycker
circa 1910

AN EYE ON FLANDERS: THE GRAPHIC ART OF JULES DE BRUYCKER

Stephen H. Goddard

The art of Jules De Bruycker is intimately interwoven with the geography and lore of his hometown of Ghent, Belgium. Some of his biographers have used this fact to attempt to align De Bruycker's art with some of the vital aspects of the city's political history. Specifically, De Bruycker has been described as sympathetic to Ghent's central role in Flemish socialism and the Flemish Movement, which championed Flemish rights and culture. Seen in that light, De Bruycker offered an ideal of a working-class Flemish artist steeped in the task of describing the poor and disenfranchised of Ghent. In addition, his satirical and whimsical bent inevitably led to comparisons with earlier Flemish masters, notably Pieter Bruegel the Elder. Today it is not uncommon to hear De Bruycker described in Belgium as "too Flemish, too Belgian." This perception of De Bruycker as something of a cultural extremist has allowed him to be implicated in chauvinistic texts or to be construed as a reactionary, even xenophobic figure. However, the truth is that he remained aloof, avoiding any particular political association and disavowing his humble origins.

De Bruycker's aloofness went beyond his political and social stance. Like many other independents in the twentieth century, he ignored the path of modernism and worked instead in his own idiosyncratic style, which dwelled tenaciously on the dignity of the old city, its traditions, and its inhabitants. His career might have been assessed as the nostalgic brooding of a malcontent had it not been interrupted twice by a world war. De Bruycker's sustained activity as an inveterate and careful observer lends his wartime prints an undeniable authority, and in retrospect his wartime prints assure us of the depth of compassion in his scenes of Ghent.

THE CITY CONTEXT

Ghent [Flemish: Gent, French: Gand], the capitol of the province of East Flanders, is central to Flemish history, geography, and culture. The origins of this city on the rivers Leie and Lieve (near the river Scheldt) go back to the establishment of two abbeys, Sint-Pieters and Sint-Baafs, in the seventh century. After the abbey of Sint-Baaf was destroyed, the chapter was transferred to the Church of Sint-Jan, which became the Cathedral of Sint-Baaf in 1559 and houses one of the masterpieces of Northern European painting, the Ghent Altarpiece by the brothers Hubert and Jan Van Eyck. Ghent is graced with several other beautiful historical buildings: the medieval Cloth Hall, Belfry, and Town Hall; the Castle of the Counts of Flanders [the Gravensteen or Gravenkasteel] whose foundations date back to the ninth century; and the churches of Sint-Jacobs (12th–15th c), Sint-Niklaas (13th–14th c), and Sint-Michiels (15th–17th c). Two beguinages [*begijnhofs*, convents for the *begijns*, members of a lay sisterhood] also were situated in Ghent beginning in the 13th century. These two *begijnhofs* had nearly a thousand Sisters in residence in 1910, about two thirds of the *begijns* then still active in Belgium.[1]

Many historical events point to Ghent's central role in the Flemish provinces. In 867–68 Baldwin Iron Arm, first Count of Flanders, built his castle in Ghent to defend against the Norsemen. The city established its wealth on a burgeoning textile industry. Ultimately the Count of Flanders, who was loyal to the French King, and the wealthy merchants in Ghent's textile industry came to an impasse, and the citizens sought, and eventually won, independence from the French. Under the leadership of Jan Borluut, soldiers from Ghent participated in the Battle of the Golden Spurs near Kortrijk in 1302, in which the best of the French Chivalry was entrapped and routed by Flemish soldiers armed with pikes.[2] Ghent found a martyr in Jacob Van Artevelde, who had fought for the interests of the merchants and sided with England (the source of wool for the Flemish weaving industry) rather than France in the early years of the Hundred Years War. Van Artevelde, who managed to form an alliance among the cities of Ghent, Bruges, and Ieper with himself as Captain General, was murdered in 1345 at an uprising incited by his plan to change Ghent's allegiance from Count Louis of Flanders to the Black Prince of Wales. Ghent achieved tremendous wealth during the years of Van Artevelde's stewardship and he has become emblematic of the city's independence; his statue of 1863 is a well-known landmark in the Vrijdagmarkt [Friday Market].

With the introduction of cotton mills in the nineteenth-century, the textile industry again became the heart of Ghent's economy. The increasing population of weavers and spinners led to myriad social problems and, in turn, to the formation of pre-socialist workers organizations. As early as 1857 the Brotherly Society of Ghent Weavers and the Society of Destitute Brothers [Broederlijke Maatschappij van Gentsche

Fig. 2. The Tichelrei, just north of the Patershol beyond the Sluizekenkaai. Courtesy Stedelijke Commissie van Monumenten en Stadsgezichten Gent.

Wevers and Maatschappij der Noodlijdende Broeders] were established in Ghent.[3] With the advocacy of Emiel Moyson, a system of medical and social assistance for the needy was created, and ultimately the workers of Ghent found mutual support through the various forms of aid sponsored by the peoples' meeting hall, the Vooruit. The socialist newspaper that takes its name from the Vooruit often published articles on the art of Jules De Bruycker.

Together with Antwerp, Ghent played a significant role in the Flemish Movement, which sought the same legal and social footing for the Flemish-speaking population of Belgium as had long been enjoyed by the French-speaking.[4] The Royal Flemish Academy for Literature and Linguistics [Koninklijke Vlaamse Academie voor Taal- en Letterkunde] was established in Ghent in 1886. Tokens of the "language problem" occasionally appear in De Bruycker's works, but as was the case with political issues in general, he did not reveal his opinions in his work. It would appear that he alienated neither linguistic camp, for among the distinguished literary figures from Ghent who wrote admiringly about De Bruycker was a leading figure in Flemish letters, Karel van de Woestijne, and three others who wrote in French: Jean de Bosschère, Grégoire le Roy and Franz Hellens.[5] Even De Bruycker's close friend Peter Bonnel emphasized the artist's essentially apolitical stance.[6] In his private correspondence with the Dutch collectors Jacob and Louise de Graaff-Bachiene, however, De Bruycker expressed a decidedly pro-Flemish attitude, and it is probably significant that in addition to his following among the readership of *Vooruit*, one of the key studies on De Bruycker was by Achilles Mussche, a leading Flemish socialist, advocate of Flemish letters, and native of Ghent.

Ghent also played an important role in art and literature.[7] The city claims considerable artistic ancestry in the fact that the van Eyck brothers and Hugo van der Goes lived and worked there in the fifteenth-century. In De Bruycker's own time some

Fig. 3. The Sint-Veerleplein prior to restoration of the Gravensteen [Castle of the Counts, at right back]. Courtesy Stedelijke Commissie van Monumenten en Stadsgezichten Gent.

of the key artistic figures in Ghent were the painter Théo van Rysselberghe, one of the central figures in the Belgian avant-garde artists' group Les XX; the sculptor Georges Minne, also of Les XX, whose work was of considerable importance in the genesis of European expressionism; the talented landscape and portrait artist Gustave van de Woestijne, who was, with Minne, central to the Sint-Martens-Latem artists' colony near Ghent; and the woodcut artist Frans Masereel (later active in Paris), whose enormous output of stark black and white images on the themes of the contradictions of modern life were often put into the service of international socialism. In the literary realm, in addition to the previously mentioned Franz Hellens, Grégoire le Roy, and Karel van de Woestijne (brother of Gustave), Ghent also claims the symbolist playwright Maurice Maeterlinck, who won the Nobel Prize for literature in 1911.

It should be borne in mind that Ghent is not unique among Belgian cities in its ability to conjure up a rich past and a vigorous modern culture. The features that distinguish Ghent are those that Achilles Mussche spelled out in rather effusive poetic terms in his discussion of De Bruycker's origins. Mussche speaks of Ghent as a city of contrasts whose working population, whether in service to the woolen industry of the Middle Ages or to the cotton mills of De Bruycker's day, lived in the poor quarters situated between the antithetical emblems of clerical and temporal power: the Abbeys and Castle of the Counts:

> In the watchtowers and chapel a cotton mill screeches and drones; men women and children are worse off than animals and toil in the falling cold. Hundreds of pitiful hovels and pubs fester like leprous bumps on the cold hard stones of the castle—an incomparable image of class contrast.[8]

Even if somewhat contrived, the stereotype of "turbulent, seething Ghent...where medieval Europeans first rose against privilege and power" or Ghent as "the bubbling cauldron of Europe" is one that its own citizens have done much to promote.[9] While De Bruycker's art has been pressed into service to foster just this image of Ghent, a careful examination will show that he had a dispassionate, bemused, and ironic view of the mundane comings and goings of his compatriots.

Although scaffolding adorns many of the buildings depicted in De Bruycker's works, he contemplated the gussying up of medieval Ghent in preparation for the 1913 World Exposition with amusement and detachment. Around the turn of the century Ghent was subject to many transformations. Sections of the Lieve River and old canals were vaulted over [fig. 2], new boulevards were constructed, and the canal system was much improved so that larger vessels could reach Ghent from the sea. Under the administration of Hippolyte Lippens (Mayor 1882–95) some of the "leprous quarters, the unhealthy enclosures where the workers were penned up," such as the Kattenberg, the Kalleitje, the Gruisberg, the Nieuwpoortje, and the alleys known as Reep and Veer, were subject to urban renewal and replaced with "well ventilated streets, spacious structures, or elegant squares."[10] The Lippens administration also supported the restoration of many public monuments, notably the Castle of the Counts [fig. 3]. Mayor Lippens was succeeded by the city engineer, Emile Braun (Mayor 1896–1907), who continued "renovations" and the erection of new streets and structures with fervor. Following the models of Baron George Haussmann in Paris and Mayor Jules Anspach in Brussels, Braun's administration oversaw the gutting of the old center of town in order that the views of the Belfry, Town Hall, Cloth Hall, Sint-Baafs, and Sint-Niklaas would be unencumbered by smaller structures.[11] While the church of Sint-Niklaas was of special significance to De Bruycker, the old streets, with names like Luizengevecht, Serpentstraat, Bloedsteeg [Louse-battle, Snake Street, Blood Alley] served as De Bruycker's primary inspiration in his early career.

CLASS CONSCIOUSNESS AND SELF-CONSCIOUSNESS

Jules De Bruycker was born March 29, 1870, in one of the old quarters of Ghent on Jan Breydelstraat, no. 9. His family ran an upholstery and wallpapering business that De Bruycker continued to manage for much of his life. The young De Bruycker displayed promise as an artist and, at age ten, began to attend art school at the Academy of Fine Art in Ghent. This study was interrupted after four years by the death of his father. For many years De Bruycker worked on household interiors to help the family, while he did his best to pursue his art on the side. Much to his annoyance, even after enjoying considerable success as an artist De Bruycker was still referred to as the *tapissier* or *behangersgast* [upholsterer or wallpaperer]. In a gesture of self-ridicule typical of De Bruycker, he added a droll depiction of himself beating wool stuffing to refurbish some furniture, inscribed "the beginning of our career," next to his signature in an impression of one of his prints [cat. 14].[12]

The degree to which De Bruycker's daytime job surfaces in discussions by his contemporaries and compatriots tells us a good deal about the level of class consciousness in Ghent. Consider, for example, the story told on several occasions by the Ghent writer Franz Hellens, author of *En Ville Morte* [In the Dead City, cat. 6 and 7], a dark and evocative novel published in 1907, steeped in the "dead city theme" established by Belgian symbolist author Georges Rodenbach's *Bruges la Morte* [Bruges the Dead].[13] Hellens, who must have been familiar with De Bruycker's drawings, realized that they would make a perfect compliment to his gloomy narrative set in old Ghent and, sometime around 1905, set out to meet the artist:

> I got his address and I decided to go knock on his door. To my great disappointment, he lived in a tidy little house on an ordinary street in the new part of town....No doubt I would find myself face to face with a poorly dressed being whose lean and sickly face and long unkempt hair immediately indicated the traits of an artist... A small well-dressed man with a conscientious air and an entirely normal likeness opened the door.
>
> Where had I seen this figure before? I remembered. My parents lived in a bourgeois house between town and country. One afternoon I was busy slogging away at the material for my next exam when someone knocked at the door. I answered in an ill humor and scarcely glanced at the intruder, an ordinary man of about thirty years, whose contours, at first, seemed devoid of character. He had a carefully trimmed short beard and held a bowler in his hand. "I am the tapissier," he said to me in a subdued voice.[14]

The figure, of course, was De Bruycker, who had come to repair a window for Hellens' parents.

De Bruycker joked about this inescapable identity; the following passage follows his description of a Ghent *tapissier* who had left the trade to make a fortune performing balancing acts:

> Am I not a failed tapissier? No doubt one learns most from bad experiences. I too dreamed of going around and somersaulting, in a modest way, of course, with my pencil—because I never had the disposition for the stage or the circus.
>
> I drew over and over, and my father put this to good advantage and sent me to the academy. He thought that the study of styles would be useful for a tapissier.
>
> We were at the dawn of the age when the bourgeois dreamed of a Renaissance dining room and a Louis XVI living room.[15]

De Bruycker further decried that even after attempting a disguise by shaving his beard following his long absence from Ghent during World War I (during which time he had enjoyed considerable success as an expatriate artist) he was greeted back home by passers-by as the *tapissier*:

> Pow! With a blow I felt the little aureole that I thought adorned me go out. More than fifty etchings, numerous watercolors, and thousands of drawings amounted to nothing for my home town. I was always *le tapissier*, the humble upholsterer![16]

Perhaps the barb was nowhere more forcefully, and probably intentionally, applied than by Karel van de Woestijne who recollects in a 1922 article, "We weren't yet artists: you were a wallpaperer [*behangersjongen*] (now may I blab it out?), I a student in the first levels of the university."[17] Both Hellens and van de Woestijne make a clear distinction between themselves—young privileged members of the intelligentsia attending the university—and De Bruycker, whom they place irrevocably in the role of a common worker. This distinction was not without irony; as we will see, De Bruycker sought respect as an artist and distance from his humble origins while the brothers van de Woestijne sought solace and inspiration from the rural peasantry in the countryside outside of Ghent.

In 1893, after nine years absence, De Bruycker re-entered the Ghent Academy, where he studied with the painters Théo Canell, Louis Tijtgadt, and Jean-Joseph Delvin. During this period and the years that followed, while he took up residence in a bohemian community, De Bruycker continued to work in the family trade. This was also when De Bruycker, who had yet to learn to make prints, concentrated on making drawings of the inhabitants of the poorer quarters of Ghent. Delvin later wrote to one of De Bruycker's collector's, René van Herrewege:

> Long ago De Bruycker once let me into the room where he worked—I don't remember how or why. I was completely overwhelmed. Everywhere drawings on little loose pieces of paper, carelessly and deliberately unfinished to entice the interested. Everything was used, in every possible way just to give the impression of an overwhelming train of thought. There I saw human wretchedness laid bare, as it is, distinct or obscure.
>
> And I remember the peaceful figure of De Bruycker. Only his fleeting eyes betrayed his restlessness. I remember his blazing glance, whenever he maintained that he had not made a caricature.
>
> "An art to make you laugh?" he mused. "My soul cries whenever I so portray humanity. I see them so, and this is how they are. I have compassion for them, endless compassion."[18]

PATERSHOL

In 1902, in his early years as an artist (though he was already over 30), De Bruycker took up residence at an abandoned Carmelite abbey in the district across from the Gravensteen called the Patershol, which can be translated roughly as "the monks' den" [figs. 4, 5]. This was the most notorious bohemian locale in turn-of-the-century Ghent. Many of the leading writers and artists of the day visited the Patershol, and it has been evocatively described in both of Karel van de Woestijne's long articles about De Bruycker. This is how De Bruycker described his new haunt:

> In 1902 I changed my quarters and set myself up in the Patershol—a largely ruined ancient convent—where more than one artist established a residence. These makeshift lodgings nearly all opened onto a vast interior court which allowed a favorable light. It was a picturesque milieu, not only for the decrepitude of the buildings, but also for the variety of the tenants: artists and poor, drunk incorrigibles. This is where I grappled with my large format watercolors.[19]

As Robert Hoozee has demonstrated, Karel van de Woestijne and his criticism were central to Flemish art in the early twentieth century.[20] Twice van de Woestijne published long articles on De Bruycker, in 1912 and 1922. Rather than in-depth studies of the artist, however, these articles made use of De Bruycker as a touchstone for considering Ghent and its environment, especially its seedy, bohemian milieu. They also offered the author opportunities to reminisce about his hometown and to recapitulate some of the seminal events and characters in turn-of-the-century Flemish arts and letters. These articles paint a vivid picture of De Bruycker's haunts and merit a careful reading. Among the many digressions, van de Woestijne fondly recalls visits by Flemish authors Stijn Streuvels and Herman Teirlinck to the Patershol, as well as his meeting with Jules de Praetere there. De Praetere was among the most gifted book designers and printers of Flemish literature. Among his achievements is his beautiful edition of Stijn Streuvels' *Lenteleven* [Spring Life]. As van de Woestijne recalled ten years later, in 1912:

> Sitting round that stove on rather rickety chairs, we smoked our clay pipes. Remember Herman Teirlinck, when we were

Fig. 4. Entry to the Carmelite Pand from the Vrouwebroersstraat (photographed in 1993).

An Eye on Flanders: The Graphic Art of Jules De Bruycker

Fig. 5. The Carmelite Pand. Courtesy Stedelijke Commissie van Monumenten en Stadsgezichten Gent.

studying in Ghent at the time, how we wrangled about Shakespearean criticism around that intimate little stove. And all the other mates, how we analyzed Kropotkin. And you, Julius De Praetere yourself, who have since become Herr Direktor of the Zurich Kunstgewerbe-Museums, remember how we glowed with enthusiasm for a new art of the book; how we, with no money and without certainty, began the publication "Werk," which, after the first series of "Van Nu en Straks" (still its most beautiful number whose printing, ornamentation, and highly logical typography rendered it the most consciously executed periodical)[21]... remember, how we (while searching, from inn to inn, for the best Oudenaard beer), lived in a restless and wavering, yet wonderful intellectual life!...

Further in the hall stood the unwieldy handpress, amidst tables and easels, the press De Praetere used to realize his insight into the art of printing. The pages of Stijn Streuvels' "Lenteleven" dried on long ropes, for the first edition was issued here.[22]

Both Karel and his brother, the painter Gustave van de Woestijne, had frequented the Patershol, but another artists' colony was more central to their activity. This was the colony of artists who gathered in the countryside at Sint-Martens-Latem, outside of Ghent. From van de Woestijne's vantage point, De Bruycker and the denizens of the Patershol represented an urban counterpoint to the immersion in rusticity that the artists of Sint-Martens-Latem pursued. This was not just a return to nature, but to some degree an attempt to return to the pristine world that could be gleaned from the paintings of the fifteenth-century "Flemish Primitives." There is an undeniable naiveté in the young van de Woestijne brothers' sojourn at Sint-Martens-Latem, and this too may be construed as a counterpoint to more worldly existence in the Patershol. Gustave wrote a book about the years at Sint-Martens-Latem called *Karel en Ik* [Karel and I]; it is a valuable and unusually frank testimony. Gustav's anecdotal text not only tells us much about the informal gatherings of the artists of Sint-Martens-Latem; we also learn that when the brothers van de Woestijne proved to be incapable of properly caring for themselves (they became sick from eating nothing but eggs) their mother arranged for a maid to look after them. They had a gardener as well. All of this afforded them the luxury of time to create some of the most remarkable paintings and texts of twentieth-century Belgium.

Karel van de Woestijne's lengthy evocation of the Patershol, by way of discussing De Bruycker, was a means of writing the history of the avant-garde that developed in the environs of Ghent. While there is a distinguishable reluctance to place De Bruycker in the center of this avant-garde, van de Woestijne's 1922 article gives one of the most intimate, recollections of the Patershol and includes a description of his meeting of De Bruycker there:

As with other artists who had put up at the Pand, the Rolleken or the Vrouwebroers, he [De Bruycker] lived in a locale of the Patershol. Pand, Rolleken, Vrouwebroers and Patershol were interwoven monastic complexes. Earlier the Pand was dependent on the masters of Saint Michael's [Sinte-Michielsheeren]; the Rolleken owed its name to the niche, divided by a partition, wherein people (an example is given by no one less than Jean-Jacques Rousseau) laid down newborn children for whom there was no place in the hearts of the mother and father. The brothers of Our Lady gave their name to the third giant building. The "Patershol" did not have a double meaning; it was an ink black hole where in previous centuries hundreds of monks ("paters" or fathers) lived. The last two localities, "Vrouwebroers" and "Patershol," were especially ruinous, but were able to live on as architectonic curiosities. They stood, bony and immovable, in the heavy shadow of the nearly thousand-year-old Gravensteen, where at the time swarms of workers were making lengthy reparations. These buildings blocked the narrow, long, winding alleys in the middle of which a trench made its smokey meander, and from the infinitely high and listing houses wafted the smell of lye and coal smoke that was the natural evaporation of the wives who could be heard quarreling and whose children could be heard whining. The Vrouwebroers shone with a beneficent cleanliness. A large gate led to a courtyard that, to the left, showed the broad leaves of tall sunflowers on the threshold of the little houses where the old people had their homes. To the right, in the clear corridor that enclosed the side of the courtyard and its grassy cobblestones, there was a wide oak stairway [fig. 6] with decorative carving that led to the cells that had been fixed up as ateliers. One usually found award-winning painters there, it was a dignified place.

The principal building in the Patershol, which lay right across from a suspicious inn that was full of deathly quiet until ten o'clock at night—and then the nightly screaming commenced as if someone had been murdered under the final command of hellish powers—the principal building in the Patershol saw its courtyard taken over by a cooper who burned a little wood fire under his fir cask that burned the eyes and throat. To the left, under the old slate roof that declined on a handsomely carved rafter, a steep stone stairway turned, with the help of a rope shining with grease, where the painters ateliers opened up. They were large, blue-plastered spaces, furnished with a singularly eloquent slovenliness. There was no single hole through which you could see the slightest decoration. On the other hand, unmade beds with dubious sheets offered a hospitable sleeping accommodation. The lack of paintings here was noteworthy: impoverished painters rarely think of work. But one noted unwashed plates here that suggested that something was eaten now and then. The walls did not yield pride of place to those of Italian palaces: there were many naked figures drawn in charcoal in a moment of megalomania. A pile of books lay under a rusty revolver: inviolable fruit of the mind. And further on there was the misery of a plaintive man.

I have begun with the worst. Among them was the cell of someone who had actually made it as far as professor in the drawing academy. He had a consumptive wife in a room in town: therefore he didn't need a bed here. He had a harmonium in its place that played the role of stimulus. Schiller surrounded himself with rotten apples, Puvis de Chavannes brushed his legs with eau de cologne, Beethoven poured water over his wrists; this figure played Gounod's *Ave Maria* on an out of tune harmonium. He needed to do this in order to paint the Stations of the Cross, which was his continual occupation. An interment was scarcely done and he would stretch a canvas for the scene of Christ Before Pontius Pilate. With the greatest regularity the other twelve stations were painted. There was never the least alteration in painting, nor in drawing nor color. We knew exactly when he was about to start an anguished look with the tip

Fig. 6. Wooden stairs in the Carmelite Pand. Courtesy Stedelijke Commissie van Monumenten en Stadsgezichten Gent.

his round brush (sign of his skill): he played an Ave Maria, and there was the inspiration to do it. We had a sort of disdainful pity for him; his wife had consumption, his sallow children continuously snotnosed.

It was under the inspirational weepiness of the Gounod harmonium that I laid my hand for the first time in that of Jules de Bruycker, upholsterer (*behangersgast*). It was in the lowest of the rooms; high windows let in the smell of the wood fire that warmed the belly of the fir barrels in which apples would be sent to England; which was at the time a Ghent occupation that I never properly understood. This spacious room with tall vaults, where the vestige of an altar seemed to be, was as good as nothing in bringing painting to mind. But there was a Gothic grammar, the English exercises by Stoffel and a critical edition of [Schiller's tragedy] *Kabale und Liebe*: my property. Shared property included a pile of numbers of *Les Temps nouveaux* and the

> *Mercure de France.* There was also a very handsome pipe, that had no other pleasure than to sop, that I had made a communal gift of. About this time this locale was honored with a visit from Stijn Streuvels and Herman Teirlinck. On this occasion the red tile floor was swept clean by a model invited for this purpose. She was a sleepy gal full of submissiveness. She thought she had been invited to pose, but it was just to sweep. For her it was a disappointment that she didn't have to disrobe. She had never accepted a cent for sweeping. ...
> In this room and under these circumstances [ruminating about painting theory, *Kabale und Liebe*] that I first met Jules de Bruycker. He was not agreeable: nearly ten years older than I, gifted with a talent already valued by his friends, but he had never held an etching needle in his hand, only upholsterers nails that he used to use—he sprang in on us, the strap of a round, red carpetbag over his shoulder. He had to care for his mother and sister, he worked with intensity and bitterness. He was shy and a skeptic; he laughed with contempt when it was said that he could earn more as a draftsman than as *tapissier*. He was old enough not to give in to idle fancy. There was something that he knew only too well: he would never be a good painter as he was as good as colorblind. He, who was already a grandiose caricaturist, with all the qualities of a Daumier, nevertheless lacked Daumier's somber colors. Sometimes he brought us drawings, and once on request he watercolored one of his drawings (it was, I think, his first commission). Then we looked at him: his clownish head distorted and trembled around the mouth. His wonderful pointed nose that really *looks*, his penetrating eyes, his small and wonderfully movable lips, his entire gaunt and supple body, they were a negation, that signified everything: nothing, nothing.
> And nevertheless he could not live without drawing. More than an obsession, it was a need with him. Meeting a type that caught his attention on the street, he followed him, a paper in the palm, pencil between his fingers. He forgot his wallpapering and upholstery work (*behangerswerk*); he was possessed by art.[23]

Van de Woestijne apparently met De Bruycker in a central, communal room, where pipe smoking and discussions of the anarchist thinker Kropotkin or the socialist journal *Les Temps nouveaux* prevailed. However, De Bruycker seems to have preferred the company of several noted eccentrics in the Patershol, among them Cies de Kalle (pseudonym for Georges vande Walle, see cat. 1) and a certain Van der Swalm, respectively characterized by Chabot as "a painter who claimed to be followed by light rays, and a former mason who presented himself as owner of the Gravenkasteel—each called the other crazy."[24] Van de Woestijne claims that de Kalle got his name (which means magpie) because of his habit of springing about on his long legs, continuously prattling in a mix of French and Flemish.[25] Van de Woestijne continued to describe de Kalle, a frequent visitor to de Praetere's studio in the Patershol, as suffering from a persecution complex and as subject to nocturnal hallucinations.

Franz Hellens later encountered De Bruycker and de Kalle in the kitchen of Emile van Vooren, an eccentric caretaker at the University.[26] Hellens made these unusual meetings at van Vooren's the subject of a short story, "La Cuisine des Fous" [The Fool's Kitchen]:

> The artist occupied a conspicuous seat in the caretaker's kitchen. Mr. Charles admired him with the naïveté of a child before a priest, although the privileged one didn't appear to care, and contented himself to find the heat excellent, the cigars fine, the place comfortable and profitable. His mobile anteater nose seemed ceaselessly to seek the sensual scent of things. In the overheated atmosphere of this kitchen, where the animal vapors of the musty grease of the

Fig. 7. Albert Baertsoen, *Pluie à Gand* (Rain in Ghent), etching and drypoint. ©Bibliothèque royale Albert 1er, Cabinet des Estampes, Brussels.

food melded with his tracker's instincts, he made thousands of studies solely with the appearance of warming himself and chatting. The walls were covered with unfinished sketches, the crazy burden of which neither the simplicity of the caretaker nor the senility of his dog respected. The caretaker undertook to sell the drawings, they were behind the cell, in a junk room, in boxes piled high with studies. Collectors came from afar! and Mr. Charles said proudly, 'I work for my artists!'[27]

It is around this time that De Bruycker discovered the art of etching and with etching he found a means of amalgamating his random sketches and studies. In 1905 he stumbled upon the etchings of the Ghent artist Albert Baertsoen in the Ghent Museum of Fine Art.[28] De Bruycker recalled, "It was a revelation! My wild enthusiasm for this process was immediate and irresistible."[29] In Baertsoen De Bruycker found precedent for his own moody images of the old quarters of Ghent [fig. 7]. De Bruycker found an adept to instruct him in the art of etching in the painter Fritz van Loo. Van Loo, who had also been a student of Jean-Joseph Delvin and Louis Tijtgadt at the Ghent Academy, resided in the old Dominican cloister, the *Pand* (not to be confused with the Carmelite *Pand* at the Patershol previously discussed).[30] De Bruycker had his plates printed in Elsene (Ixelles in French, a quarter in Brussels) by the Van Campenhout firm and they were often published by the distinguished fine arts bookstore Maison Dietrich, also in Brussels.[31]

THE EARLY ETCHINGS

De Bruycker's early etchings, dating from 1906 until the outbreak of the First World War, concern four primary themes: the artist's haunts in and around the Patershol; the open-air flea markets with their rough, low-life population; the cheap sections of the theater; and some of the historic sites of old Ghent. De Bruycker unfailingly shows us these sites swarming with urchins and the working poor. His etching and drypoint *Confrère* [Colleague, cat. 1], 1906, for example, shows the lanky Cies de Kalle drawing in a den, probably in the Patershol. Many of the street and market scenes can still be located today.[32] *Ruelle (Gand)* [Alley (Ghent), cat. 2] and *Ruelle (Patershol)* [Alley (Patershol), cat. 4] may both look down Haringsteeg, from slightly different vantage points near the intersection with Ballenstraat. The gutter in the center of the street, probably an open sewer in De Bruycker's time, leads our eye to the ominous (and in *Ruelle (Gand)* exaggerated) hulk of the antithetical Castle of the Counts. In *Veergrepe* [cat. 5, visible on the wall of De Bruycker's studio in his portrait of Frans Masereel of 1909, fig. 8] we see a courtyard behind shabby dwellings from across the river Leie.[33] The streetside access is just visible in the upper left. A period photograph shows that De Bruycker did not bother to compensate for the fact that the image would be reversed when printed [fig. 9]. Although this view of washing and morning chores is clearly not contrived, De Bruycker's impulse for amused or purposeful exaggeration places him solidly in the realm of caricature, however stringently he was to reject this notion. We can imagine that De Bruycker was ill at ease with the term "caricaturist" because it may well have militated against his striving

Fig. 8. Jules De Bruycker, *Frans Masereel*, watercolor, 1909. Museum voor Schone Kunsten, Ghent.

Fig. 9. The Veergrepe. Courtesy Stedelijke Commissie van Monumenten en Stadsgezichten Gent.

for respect as an artist in the eyes of the bourgeois Belgian public who were already somewhat ill-disposed to printmaking as an art form, let alone to caricature.[34] De Bruycker is to Ghent perhaps what the slightly older and more emphatically political artist Heinrich Zille was to Berlin. Both artists give us more intimate glimpses of daily life in early modern Europe than was afforded by many of their colleagues who were more in step with the general contours of modernism. Within a few pages Karel van de Woestijne found it possible to speak of De Bruycker's art in terms of sarcasm, pessimism, nihilism, skepticism, irony, and caricature.[35] To a devout figure such as van de Woestijne (who, for example, had brought the Sint-Martens-Latem painter Valerius de Sadeleer back into the arms of the church following the latter's confession of agnosticism and sympathy for anarchism), De Bruycker's wry and unorthodox renderings may have seemed to smack of lack of faith.[36]

That De Bruycker was sensitive to the ironies of depicting "picturesque misery" (to use Mussche's phrase) is made clear by his 1907 etching, *Rolweg Brugge* [The Rolweg in Bruges, cat. 14].[37] In this unforgettable image De Bruycker mocks artists who create considerable clutter in an alley in a poor neighborhood in Bruges with all their gear and subject the local poor to their scrutiny. They focus all their attention on the emaciated mother and child and the urchins in rags who seem to be outnumbered by artists. De Bruycker does not hesitate to ridicule himself in this scene; he is the lanky artist at the left, while his heavyset colleague seen from the back is no less than Valerius de Sadeleer, one of the finest painters of the Sint-Martens-Latem group.[38]

The open air markets and cheap sections of the theater (called "Paradise" or "Owls' Roost" since they are high up in the theater, see cat. 8, 9, 11) were among De Bruycker's most fertile arenas for studying his fellow citizens of Ghent [Gentenaars].[39] His method must have been to compile numerous quick sketches (probably discretely on bits of paper concealed in his hand), such as those in the volume of mounted studies in the Royal Library in Brussels [figs. 10, 11].[40] These loose studies could then be worked into larger drawings and etchings. As with his street scenes, the topographical description of the market scenes is fairly accurate. The settings for the fish market at the Sint-Veerleplein (where, incidentally, Jules de Praetere's father was the auctioneer in the fish market) [cat. 3, compare with fig. 3] and the markets set up on the paved area surrounding the church of Sint-Jacobs [cat. 10, 12, 13] are readily recognizable today, although De Bruycker took some liberties in depicting these scenes.[41] For example, in *Jour de Marché à Gand* the verticality of

Fig. 10. Jules De Bruycker, *Étude–marché* [Study–Market], graphite on paper. ©Bibliothèque royale Albert 1er, Cabinet des Estampes, Brussels.

Fig. 11. Jules De Bruycker, *Étude* [Study], graphite and watercolor on paper. ©Bibliothèque royale Albert 1er, Cabinet des Estampes, Brussels.

the tower of the Castle of the Counts in the background, which is seen under restoration, has been exaggerated.

Karel van de Woestijne gives considerable attention to De Bruycker's market scenes, doubtless recalling them from his own first-hand experience. He lists with relish the variety of humanity and rummage visible in one such composition:

> Long greedy fingers weigh strange keys, they lift up iron roasters in which rust has eaten decorative arabesques, they handle fire pokers from Damascus, or simply a pair of not-too-corroded flat irons, or else a rare set of eighteenth-century novels, or sixteenth-century pamphlets bound in gray, the complete works of Buffon or a collection of fashion prints, or perhaps a holy relic, a bed pan, an imperial portrait, a dishcloth to polish the stove, civic guard equipment, a lamp, a revolver, a medieval town ledger, old porcelain, worn out clothes (about which there are many questions), musical instruments or a can of prehistoric shoe polish.[42]

However, after reveling in the multiplicity of the market scenes, van de Woestijne returns to them some pages later in order to elevate them as a form of "holy image," a "beautiful image of inner sympathy," evoked by the simple figures whose old hands have worked so hard and asked for so little.[43] This empathy is certainly something De Bruycker was aiming for, although it is unlikely that he would have ever considered himself a maker of holy images.

Civic monuments also figure in De Bruycker's pre-war etchings, but as usual they writhe with humanity on the street level. Among his most admired etchings of Ghent landmarks is his rendering of the beautiful baroque building *De Fluitspeler* [The Fluteplayer] or *Het Vliegend Hert* [The Flying Deer], home of the noted early eighteenth-century anatomist Jan Palfijn. This 1669 building is shown with the adjacent house *De werken van barmhartigheid* [The Works of Mercy], also of the seventeenth-century. Both buildings are ornamented with relief carvings, those on *De Fluitspeler* are in terra cotta (the cartouche in the gable shows a fluteplayer).[44] These homes are nos. 77 and 79 on the Kraanlei, in the immediate vicinity of the Sint-Veerleplein, the

Fig. 12. Brussels, *Palais de Justice*, chromolithograph. Private collection.

Castle of the Counts, and the Patershol. The buildings' elegant baroque gables and friezes contrast with a hurly-burly street scene below [cat. 15], much as in De Bruycker's etching of the Castle of the Counts [*Rond het s'Graven Kasteel te Gent*, cat. 16] the medieval mountain of masonry contrasts with all sorts of Rabelaisian activities. Nowhere is De Bruycker closer to Goya than in *Rond het s'Graven Kasteel te Gent*. In fact the central figure on the platform wagging something seemingly disrespectfully in the face of the monolithic testimony to time and history against which he is silhouetted might have been a conscious tribute to Goya's *caprichos*, or perhaps to his *Bobalicón*, the grinning giant who dances and clacks his castanets.

We know that De Bruycker was especially amused by the ironies inspired by the contrast of pretentious buildings, with all they signify, and their mundane human occupants. When speaking to Albert Guislain in the early 1930s about his intentions to do some work in Brussels, De Bruycker naturally gravitated to a discussion of the immense Law Courts, the *Palais de Justice* (fig. 12). Guislain had published a book on the *Palais de Justice*, which he called "the Mammoth."[45] In fact the *Palais* is among the largest buildings erected in the nineteenth century. To understand De Bruycker's anecdote that follows it helps to remember that the enormous edifice is erected on the site of earlier gallows and that it overlooks the *Marolles*, the poorer quarters in Brussels:

> I worked there. Not enough. I would love to have drawn the *Palais de Justice*. You called it the Mammoth? The terraced architecture of the Mammoth has always tempted me. It is magnificent, this immense temple steeped in light. This piling up of forms, the terracing of stones and balconies, what splendor! What a thing to do, as we say. I went there. I searched. The good people of the rue haute and the rue des minimes [in the Marolles] offered me the hospitality of their lofts and mansards. But luck was not with me. There was always an obstacle between me and your *Palais de Justice*. Regrettable! And the outer hall! What an etching! What

Fig. 13. Frank Brangwyn, *Oud huizen, Gent* (Old Houses, Ghent), etching, 1906. Brugge Stedelijke Musea, ©ACL.

a drawing! My wife and I managed, quite by accident one day, to get into a courtroom. A case was being tried. A housewife was disputing her butcher's bill. She claimed that she was given too much bone and fat, and not enough meat. Life is droll. And in this formidable basilica such a little comedy. It was human, too human.[46]

However, sometimes De Bruycker's forcing together of the anecdotal and the sublime weighed more heavily on the anecdotal side, resulting in an image whose authority is undercut, but whose ability to amuse is not. This is the case with several etchings of 1914 concerning renovations on the Ghent Belfry, such as *De man van t'belfort* (The Man from the Belfry (cat. 17) and even De Bruycker's much acclaimed *La montage du Dragon sur le beffroi de Gand* (The Mounting of the Dragon atop the Ghent Belfry, cat. 18). In the former, workers on a skimpy scaffold teetering over Ghent are buzzed by a monoplane. The figure in the foreground is De Bruycker himself laboring away on the composition, and the figure mounting the ladder with open mouth is De Bruycker's patron, the architect René van Herrewege.[47] The *Montage du Dragon* shows the fervor surrounding the events of January 8, 1913, when the renovated dragon was hoisted back up to the pinnacle of the belfry in preparation for the World's Fair.[48] De Bruycker anticipated that this etching would even surpass his highly regarded *Rond het s'Graven Kasteel te Gent*. On January 21, 1914, he wrote to René van Herrewege:

> The first state is almost finished and I expect a result superior to the *Château des comtes* (*Rond het s'Graven Kasteel te Gent*). Excuse my vanity! I have been scraping like a madman without a break for two weeks. And I can have this hope! I am biting it right now. I am proceeding differently, slowly, with an etch of six to twelve hours. This etching must be less material, more *vision* that the *Château des comtes*. I repeat, I have much hope and tenacity or obstinacy and when I weaken, I descend to my wine cellar and open one of those bottles ... not of nitric acid![49]

In fact, *Montage du Dragon* was a great success, both in Ghent and in the 1914 Venice Biënnale. Chabot reports that German and Italian firms sought the rights for reproducing the image, and that the Emperor Victor Emmanuël purchased a complete set of De Bruycker's exhibited etchings.[50]

In the large plates discussed above, *La Maison Jean palfijn*, *Rond het s'Graven Kasteel te Gent* and *Montage du Dragon*, De Bruycker achieved considerable virtuosity as an etcher; these plates are rich and chiaroscural, and though complex they stop short of being fussy. In many respects they resemble the etchings of the Anglo-Belgian artist Frank Brangwyn (fig. 13), who was to prove a valuable friend during De Bruycker's years in London during the First World War.

THE GREAT WAR

With the outbreak of war, De Bruycker joined a large expatriate Belgian population in London. Other Belgian artists relocated in England included Albert Baertsoen, Emile Claus, Hippolyte Daeye, Albert Delstanche, Valerius de Sadeleer, Emile Fabry, Charles Mertens, George Minne, Isidore Opsomer, Gustave-Max Stevens, and Gustave van de Woestijne.[51] Among this community of Belgians in exile De Bruycker met and married Raphaëlle De Leyn, also a native of Ghent.[52]

De Bruycker took up residence at Fitzgerald Avenue 9 and set up shop at Fitzroy Street 8, where Whistler had moved his studio in 1896.[53] In a letter published by Chabot, De Bruycker complained of the changes imposed on his usual mode of work:

> Here one had to stay in the studio—it was forbidden to work in the street. If you looked too earnestly at a bridge, a statue, or building, you would be suspect! And as if that weren't enough, there was the charming, foggy, rainy climate and formidable distances. It wasn't easy.[54]

Nonetheless, De Bruycker managed to produce a few plates of London scenes (cat. 23), but they were the exception.[55] As Chabot, Karel van de Woestijne, and the artist himself noted, this imposed internalization and the artist's distance from the actual events of the war resulted in unanticipated and novel works. As van de Woestijne put it, while in England De Bruycker:

> mostly lived in a hallucinatory atmosphere, this distance expanded the events, lent them a synthetic-symbolic significance that we, remaining in the country, did not yet understand.[56]

Or in De Bruycker's own words:

> I worked well enough. I had, without hearing a shot, made drawings in keeping with the war. Strange works that opened a new horizon for me in the future.[57]

Of course the war had an enormous and complex impact upon the arts (the Imperial War Museum in London has over 5000 works by artists among the allied forces alone).[58] It is remarkable to consider the breadth of reaction to the war. While De Bruycker was at work on his large visionary but traditional etchings in London, the Dadaists were (by 1916) living out their first wave of revolutionary performances and utterances in Zurich. Ghent-educated woodcut artist Frans Masereel (fig. 8), a friend of De Bruycker's, spent the war years in Switzerland to avoid serving in the Belgian military. While in Switzerland Masereel "became noted for stark woodcuts with pacifist and leftist themes, published in new, pacifist Swiss journals such as *La Feuille* in Geneva."[59] Many artist from all camps were, at the outset, enthusiastic and optimistic about the war, but through personal experiences with the horrors of battle they were quickly disillusioned or battle-scarred in more profound ways. August Macke, Franz Marc, Antonio Sant'Elia, and Umberto Boccioni were killed in battle. Raymond Duchamp-Villon and Roger de La Fresnaye died of war-related illnesses. Georges

Fig. 14. Frank Brangwyn, *Het wrak van het schip "Britannia"* [The Last of H.M.S. Britannia], etching, 1917. Brugge Stedelijke Musea, ©ACL.

Braque, Guillaume Apollinaire, and Oskar Kokoschka were wounded, and Ernst Ludwig Kirchner suffered from nervous disorders resulting from the war.[60]

The war can be seen as a crucible in which pre-war impulses, the authority of intense personal experience, and various reactions to war and the specific social and political evils it exposed interacted and propelled the arts along the main trajectories of modernism. While De Bruycker participated in this arena, it was in a personal and idiosyncratic way. His earlier immersion in the microcosm of Ghent was superseded by apocalyptic images of war infused with Flemish cultural lore, frequently of a Bruegelian caste. This upwelling seems like a rallying of a cultural genotype in the face of threatened extinction. De Bruycker was not alone in turning to popular elements of his native culture in time of war; Kasimir Malevich's wartime images made reference to Russian "Lubok" folk prints, just as Raoul Dufy's made reference to French "Images d'Épinal" folk prints.[61] De Bruycker's startling wartime images yielded to a re-immersion in the microcosm of Ghent after the war, sometimes with a lingering visionary edge. Ultimately, De Bruycker's apocalyptic reaction to the Great War as he saw it from afar proved to be quite different from his reaction to the second German occupation of Belgium of 1940, which he witnessed in person.

Nineteen sixteen was a year of explosive work for De Bruycker, as seven of his eleven images of war themes were produced and an eighth appeared in its first state. These plates were very large, all but two of them over a half meter in one dimension. Two letters of 1916 from Frank Brangwyn to De Bruycker (both in a private collection) set the tone for this production. Brangwyn was primarily involved in making propagandistic lithographs and posters during the war and some of his etchings from this period were enormous, over a meter in one dimension (fig. 14). The first letter is dated 13 August and concerns De Bruycker's need for large acid baths:

Dear Sir,

Mr. Lambotte has asked me to send you a large bath—I will let you have it early this week as the large baths I have are made of wood this hot weather has made them leak so I hope you will pardon the delay in sending it. I hope it will aid your purpose and that you will let me have the great pleasure of seeing the etchings you are now working upon.

I have long admired your work. It is splendid. I am glad to hear that you are producing plates inspired by the War. With all good wishes

Believe me my dear sir,
Frank Brangwyn

And the second of 12 September:

Dear Sir,

I had written you a letter last week which I forgot to post and which I now enclose as it contains the names of printers.

I think after seeing your proofs that Golding & Co. No. 2 on the list will be the right man for you.

I must congratulate you on the noble and fine plates you have made more especially I admire the large one of Death ringing the bell—It is splendid and if printed on better paper and by a good printer it will be splendid.

I have taken the liberty of giving your address to an American Library which is collecting plates and lithographs connected with the war. I hope you will hear from them.

With all good wishes

Yours sincerely,
Frank Brangwyn

p.s. Regarding a publisher I will give you a letter of introduction to one or two, but as you know the taste in England is more for the pretty than for the noble and large things in art.

"Death ringing the bell" refers to De Bruycker's masterpiece among the wartime plates, *Weer klepte de Dood over Vlaanderenland* (Again Death Tolled over Flanders, cat. 19). Certainly the prospects for Belgium must have looked grim in 1916, by which time two of the three battles of Ieper had been fought, establishing the arc of the Salient, the demarkation of the allied front and the locus of horrific trench and ultimately chemical warfare. In De Bruycker's print, Death appears as an enormous skeleton in hobnailed boots who, sitting astride the masonry towers of a church, has plucked the bell from the belfry and rings with it a death knell over Flanders.[62] Below in the wintry Bruegelian snowscape survivors arrive in a long funeral march laden down with caskets in a passage that, with the very notion of Death ringing a bell, recalls Bruegel's *Triumph of Death* (Madrid, Prado). Death has strung up the church's priest, whose demonic replacement and incense-toting assistant arrive from below.

In 1917 an exhibition of Belgian art was held in Rotterdam. De Bruycker's work figured prominently and the reviewer of the exhibition for the *Nieuwe Rotterdamsche Courant* singled out the image of the Death of Flanders. For this reviewer De Bruycker's print evoked not only Bosch and Bruegel, but also the figure of Ulenspiegel (usually translated as "owl glass," a heroic rogue and prankster). The legend of Ulenspiegel, originally an early sixteenth-century German text, was readily adapted and re-printed in the Low Countries. In the editions printed in Bruges and

Ghent Ulenspiegel's place of birth and/or death is given in various Flemish cities. Ultimately the text served as a point of departure for the Belgian author Charles De Coster in the mid-nineteenth century. De Coster's *La légende et les aventures héroïques, joyeuses et glorieuses d'Ulenspiegel et de Lamme Goedzak au pays de Flandres et ailleurs* (The Legend and Heroic, Joyous, and Glorious Adventures of Ulenspiegel and Lamme Goedzak in Flanders and Elsewhere) of 1867, although written in French, is an important text for the Flemish Movement because it helped solidify a sense of Flemish identity in the newly formed Belgian state. De Coster's novel, set in the sixteenth-century, heroicizes the base but clever Ulenspiegel at the expense of the clergy and temporal powers—and it allows a witty Flemish peasant to triumph over considerable odds, specifically another occupying army in Flemish history, the army of the Duke of Alba serving Philip II of Spain.[63] It is this Ulenspiegel, the Ulenspiegel of De Coster, to which the reviewer for the *Nieuwe Rotterdamsche Courant* refers.[64]

Evocations of earlier Flemish culture, from the "spirit of Ulenspiegel," to the paintings of the fifteenth-century "Flemish Primitives," abound in the De Bruycker literature. This enthusiasm for legitimizing Flemish heritage through the achievements of the fifteenth and sixteenth-centuries was not limited to De Bruycker or his apologists—it was a widespread phenomenon that found extensive expression, for example, in the nineteenth-century Flemish Renaissance Revival in architecture.[65] Belgian symbolist artists and writers had previously drawn upon the precedent of early Flemish painters, having been especially attracted to the simultaneously spiritual and realistic qualities of their paintings.[66] For example, symbolist artist Fernand Khnopff lectured on the art of van Eyck, Memling, and Metsys, and symbolist writer Maurice Maeterlinck found inspiration for one of his most important plays, *Les Aveugles* (The Blind), in a painting by Pieter Bruegel of *The Blind Leading the Blind* (Naples, Museo-Gallerie Nazionale di Capodimonte).[67]

The fervor for early Flemish art received its most significant impetus with the remarkable 1902 Bruges exhibition *Les Primitifs Flamands* (The Flemish Primitives). This exhibition inspired a lengthy article by van de Woestijne, who described himself as having been an adherent of Ruysbroeck mysticism in his Patershol and early Sint-Martens-Latem days.[68] Jan van Ruusbroec (Ruusbroek or Ruysbroeck) was a fourteenth-century mystic of Brabant, active at the hermitage in Groenendael near Brussels, whose teachings are often considered an important precursor to the Brothers of the Common Life. There are reports of van de Woestijne reading from Ruysbroeck's writings to his friends (presumably Valerius De Sadeleer, George Minne, Gustave van de Woestijne, and others) in Sint-Martens-Latem.[69] Another Flemish author, Stijn Streuvels, evoked Bruegel and Ulenspiegel in the diary he kept during the War, usually to seek comparison with some horrendous event.[70] References to the Flemish Primitives themselves were almost second nature in the circles frequented by De Bruycker. De Bruycker once remarked that he understood Rogier van der Weyden better after seeing a nun in the Sint-Niklaaskerk with her broad, white headdress.[71] Karel van de Woestijne took pride in pointing out the presumed birth place of "Flemish Primitive" Petrus Christus around Baarle near Sint-Martens-Latem; Gustave van de Woestijne described Mrs. Minne as seeming to come right out of a Flemish Primitive painting, and both De Bruycker and Karel van de Woestijne had reproductions of paintings by the "primitives" in their work areas.[72] We can see van

Fig. 15. Jules De Bruycker, *Sites et visions 3, D'une fenêtre* [Sites and Visions 3, from a Window], etching, 1932. Harvard University Art Museums M10982.

Eyck's famous portrait, *The Man with a Pink* (Staatliche Museen zu Berlin), in an exaggerated view of De Bruycker's studio in the Pand in a print of 1932 (fig. 15). Gustave van de Woestijne gives an intriguing description of his brother's writing area in Sint-Martens-Latem, which included James Ensor's etching *The Entry of Christ into Brussels*; a drawing by George Minne of Jesus with the Chalice; a portrait of the poet and editor of the Dutch literary journal *De Nieuwe Gids* (The New Guide), Willem Kloos; and a reproduction of Hugo van der Goes' late fifteenth-century *Portinari Altarpiece*.[73] Gustave's own striking paintings of the Latem peasantry also bear comparison to the precedent of van der Goes.

The Belgian enthusiasm for the Flemish Renaissance serves to introduce Jean de Bosschère, one of De Bruycker's critics, who also left Belgium to wait out the war years, primarily in London. De Bosschère had many talents: he was an author and critic who spent his later years in Paris in the circle of Antonin Artaud and André Saurès; he was a talented artist who illustrated many of his own literary texts in styles that can be loosely associated with the graphic art of Aubrey Beardsley (fig. 16) and later the surrealists; he was an intimate and important correspondent with Belgian poet Max Elskamp; and, what is germane to this discussion, he authored several early texts on various aspects of fifteenth and sixteenth-century Flemish painting and sculpture.[74] These include a monograph on the Antwerp Renaissance painter Quentin Metsys (1907), one of the first studies of Renaissance sculpture in Antwerp (1909), and a long article on Bruegel and "our taste in painting" for the Parisian review *L'occident* (1913).[75] De Bosschère reviewed De Bruycker's works in two wartime group exhibitions in London and he penned a longer article on the artist in 1918.[76] At this time De Bosschère was still fairly steeped in a Walter Pater-like estheticism. He took pleasure in describing the etcher's trade as resembling that of a tinker and he evoked

Fig. 16. Jean de Bosschère, *Le pêcheur nocturne* (Nocturnal Fisherman), photomechanical reproduction printed in blue on page 57 of de Bosschère's *Béâle-Gryne* (Paris: Bibliothèque de 'l'Occident,' 1917). Private collection.

the French etcher Charles Cotett's description of biting the plate as "delectable cookery! [cuisine delectable!]." De Bosschère conjures up an image of the etcher as a sort of necromancer, lost to the real world "like those who seek the philosopher's stone."[77] Without making reference to Bosch or Bruegel ("great names cast a shadow," De Bosschère notes), De Bosschère finds in De Bruycker a remarkable visionary, especially in the great wartime prints such as *Weer klepte de Dood over Vlaanderenland* and *Kultur!* (cat. 19, 20), and not simply someone who makes clever use of the fantastic.

Although De Bosschère nowhere spouts a nationalistic or fervently Flemish phrase (he was, after all, francophone, despite his upbringing in Lier, near Antwerp), it is significant that during his wartime stay in London he produced his lovely books, *Christmas Tales of Flanders* (1917) and *Beasts and Men* (1918, a collection of Flemish folk tales).[78] Both works were extensively illustrated by the author. As with De Bruycker, De Bosschère's drawings have been related to the spirit of Ulenspiegel, and described as satirical and grotesque.[79] Although we do not know exactly what works De Bruycker refers to (probably the illustrations for the two books of Flemish tales), we know that he admired De Bosschère's watercolors when he saw them later in an exhibition in London. He described them as "extraordinarily interesting. Flemish proverbs and tales so nicely interpreted!"[80] While in general their careers are highly dissimilar, during the war years De Bruycker and his critic De Bosschère had a brief convergence in which they mustered strength and working material by evoking their cultural heritage in the face of war.

However much we detect Bruegel in the clattering hoards of skeletons, in long-legged reapers, or simply in the theme of the Triumph of Death, De Bruycker's war images are also works of startling originality. In addition to the unforgettable image of Death over Flanders, De Bruycker invented a parade of death ironically titled, in German, *Kultur!* ("Civilization!," cat. 20), which is dominated by a cannon, a behemoth of enormous proportions straddled by an invincible skeleton who stares, unscathed, down the smoking barrel. De Bruycker's friend Peter Bonnel (who occasionally posed for the figures of fallen soldiers in De Bruycker's wartime etchings) assisted in the etching of this enormous plate. The only plate of adequate size that could be located was zinc. De Bruycker was only familiar with the technique of etching copper, and the plate was nearly destroyed when the acid-resistant ground started to lift in the acid bath. The two immediately plunged the plate in water to halt the effect of the acid, leaving only a lightly etched plate, which may explain why *Kultur!* could only sustain a small edition.[81] In *The Harvest* (*La Moisson*, cat. 21) Death and the Kaiser look on approvingly at the good labor of the grim reaper. Exquisitely cynical, Death chews on a shaft of wheat whose fecund head, silhouetted against the burning sun, offers the promise of future harvests. The outsized face that stares out from the dark of the windmill seems to be a direct quotation from Bruegel's *Dulle Griet* (Antwerp, Museum Mayer van den Berghe), but it is probably just the remains of an abandoned composition.

De Bruycker must have agonized over the grisly reports of trench warfare, which he set down in his *La Tranchée* (The Trench, cat. 22), but in *De Slechte Maere* (The Grim Reaper, cat. 24) we encounter an even more sinister aspect of the Great War, the use of chemical warfare. The etching *De Slechte Maere* also bears the titles *De dood van Ieper* (The Death of Ieper); *Ieperen de slechte maere* (Ieper, the Grim Reaper), *De Speelman Kwam Langs Ieper* (The Minstrel came along Ieper); and *De ongeluksvogel* (The Unlucky). In the background is the great conflagration that consumed the medieval Ieper cloth hall following heavy bombardment on November 22, 1914, not long before Ieper was totally destroyed (fig. 17). De Bruycker has conflated the 1914 attack with the second of the three battles of Ieper, in which the Germans used chlorine gas. This is indicated by the gas mask and bomb labeled "GAZ" in the lower

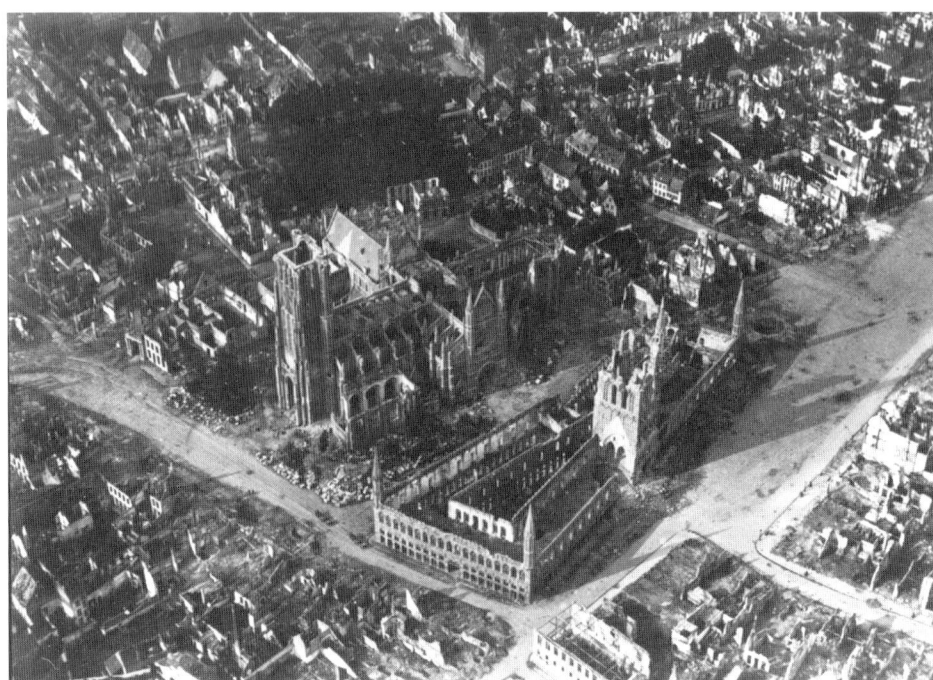

Fig. 17. The Ieper Cloth Hall and Cathedral on July 19, 1915. ©Stedelijk Museum Ieper.

left, and perhaps by the demon who is perched on the Reaper's scythe and is about to hurl a bomb. A disillusioned German soldier and writer, Rudolph Binding, described the impending fall of Ieper:

> The effects of the successful gas attack were horrible. I am not pleased with the idea of poisoning men. Of course, the entire world will rage about it first and then imitate us. All the dead lie on their backs, with clenched fists; the whole field is yellow. They say that Ieper must fall now. One can see it burning—not without a pang for the beautiful city. Langemarck is a heap of rubbish."[82]

The focus of the composition is the striding figure of the Reaper, dragging his huge scythe and with a clock surmounted by a skull on his back. This Death Clock may be a sort of rebus for the Flemish word "doodsklok," which means "death knell." The reaper's entourage includes assorted demonic creatures in a clutter of German helmets. The figure bringing up the rear carries several crowns in his cowl and seems to have an ermine robe, the dark tufts appearing in the form of demons. The overall form of this demonic emperor, as well as the remarkable lankiness of the main figure, can both find formal analogues in Bruegel's *Temptation of St. Anthony* in the Brussels museum.

Here we can return briefly to De Coster's *Ulenspiegel*. In book three, Ulenspiegel arrives at the cathedral in Ieper to recruit soldiers from the local population to fight against Alba with "les Gueux" (the opposition, who had accepted the derogatory term "the beggars" as honorific). Significantly, when De Bruycker later designed a headpiece for this chapter of De Coster's novel he returned to the essential features of *Weer klepte de Dood over Vlaanderenland* and *De Slechte Maere*: Death, with a scythe on his back, tolls the death knell while in the background a stream of figures pours forth from a church portal, another crowd carries gallows on their backs, and the reaper goes about his task (fig. 18). The specific wartime imagery involving Ieper in

De Coster's text and De Bruycker's later return to his wartime composition in illustrating this text suggest that De Bruycker may have had *Ulenspiegel* in mind when he first worked out the compositions of the Grim Reaper and of Death over Flanders.

The exhibited impression of *De Slechte Maere* was printed by Goulding Ltd. (recommended to De Bruycker by Brangwyn—the firm takes its name from Frederick Goulding, one of Whistler's printers).[83] It was dedicated to the artist's wife, twice inscribed, and after the title, the dates "1914–1918 1939–1940" have been added, in clear allusion to the second occupation of Belgium by German forces.

Fig. 18. Jules De Bruycker, headpiece for Book Three of Charles de Coster's *La légende et les aventures héroïques, joyeuses et glorieuses d'Ulenspiegel et de Lamme Goedzak au pays de Flandres et ailleurs* [The Legend and Heroic, Joyous and Glorious Adventures of Ulenspiegel and Lamme Goedzak in Flanders and Elsewhere], wood engraving, 1922. J. De Bruycker Collection.

Fig. 19. Jules De Bruycker, *Patershol artiste*, etching and aquatint, 1919. Museum voor Schone Kunsten, Ghent, 1932 CJ.

RECOGNITION AND THE SECOND WORLD WAR

De Bruycker's return to Ghent in 1919 seems to have been bittersweet. In London he had not only exhibited at the Imperial War Museum in South Kensington, but the director of the Victoria and Albert Museum had tried to convince the artist to remain in England after the war.[84] As we have seen, De Bruycker was chagrined to be welcomed back in Ghent as the *tapissier*, but it was not long before he achieved considerable recognition. He was named a Knight in the Order of Léopold in 1921; he had an important (and successful) exhibition of 154 works at the Galerie Giroux in Brussels in 1922, and in this year his work was exhibited at the Chicago Art Institute and elsewhere in the United States; in 1923 the Belgian State purchased two of his drawings; he was sought out for a teaching post at the *Hoger Instituut voor Schone Kunsten* (Higher Institute of Fine Art) in Antwerp, which he reluctantly accepted, and he was named a correspondent of the Royal Academy; in 1925 he became a full member of the Royal Academy; and in 1927 he received the national prize for fine art.[85]

The post-war years, which saw the arrival of a second daughter in 1919 (a first having been born before the war from an earlier union), were a period of considerable adjustment for the artist. De Bruycker was not at home in the first studio he found, but was much happier with one that he rented late in 1921 in the Dominican Pand (a setting that occasionally appears in the later etchings).[86] This converted cell must have resembled his old haunt in the Patershol and, like the cloister at the Patershol, a number of other artists also gravitated to the Dominican Pand between the World Wars.[87] Nineteen twenty-one is also the year in which De Bruycker began a remarkable series of etchings and drawings of craggy old figures from Ghent, notably of a certain Jacobus Alijn (cat. 25, 27) and, a few years later, of knitters and tailors (cat. 28, 29), and ultimately of himself (cat. 31, 41, 45). The 1921 etchings and drawings of Jacobus Alijn were anticipated only by the etched portrait of De Bruycker's old friend de Kalle, *Patershol artiste*, of 1919 (fig. 19). These sinewy but dignified figures, including de Kalle and the artist himself, are survivors and by the authority of their age and their evident life of hard work they testify to the weight of tradition as much as do De Bruycker's scenes of the old city. Taken together these studies as well as his return to some of his previous subjects, such as the markets of Flanders, convey a sense of closure that bridges De Bruycker's absence during the war, and that allows him to affirm the essentially unchanging qualities of his cultural landscape.[88]

De Bruycker all but stopped making prints from 1922–1924. 1922, however, saw the completion of a new edition of Charles De Coster's *Ulenspiegel* for which he contributed designs for a series of woodcut illustrations (fig. 18; cat. 26a, 26b).[89] De Bruycker must have been at home with this commission. Chabot notes that one of the editors, R. Sand of Brussels, was also organizer of the Graphics Salon (*Grafisch Salon*) in which De Bruycker had often participated.[90]

As we have seen, De Bruycker himself had been associated with the Ulenspiegel theme early on. In his 1909 text "La Cuisine des Fous" (The Fool's Kitchen), Franz Hellens evoked not only the figure of de Kalle discussed above, he also described an artist (further described as "the caricaturist") with the nickname "Ulenspiegel"—certainly this was De Bruycker.[91] The Ulenspiegel theme was everywhere visible in Belgium in the form of broadsheets, periodicals, illustrated editions, and Flemish translations of De Coster's text, as well as in musical works.[92]

De Bruycker's illustrations for *Ulenspiegel* are important because they allow us to identify certain recurrent pictorial themes encountered elsewhere in his oeuvre specifically with the Ulenspiegel theme. This is the case with the headpiece for Book III and *de Slechte Maere* already discussed. Another of the illustrations for *Ulenspiegel* looks forward to a composition worked out in more detail some six years later. The frontispiece of volume 2 of De Coster's *Ulenspiegel* (cat. 26b) is an early formulation of *Kermesse*, of 1928 (cat. 34). Also titled *Mendiants* (Beggars), *Kermesse* has become so mannered that it is difficult to read. A giant vagabond with two instruments associated traditionally with lust and low life, the bagpipes and accordion, leads a tangle of beggars beset with tokens of gambling, superstition, and religious devotion.[93] As in the book illustration, a caged owl, emblematic of Ulenspiegel, hangs near the vagabond's elbow. We can enumerate further details but their cumulative effect does little to help our understanding of the image. It might be argued that this is an intentionally enigmatic restatement of Bosch's provocative painting of a wanderer (variously titled *the Peddler*, *the Landloper*, and *Everyman*, Rotterdam, Museum Boymans-van Beuningen), which has proven to be highly resistant to interpretation. It must be conceded, however, that in some of De Bruycker's mature works, such as *Kermesse*, in which he seems to try to revive the unleashing of imagination rooted in Flemish lore that the First World War brought on, he succeeds primarily in parodying himself, giving some credence to Marijnissen's quip that De Bruycker is best when he is least himself.[94]

De Bruycker's later architectural studies include a group of large and imposing etchings whose core is a series Belgian and French cathedrals, and a group of much more intimate views, primarily of old Ghent. We have already seen that De Bruycker once spoke of his wish to etch the ostentatious Palais de Justice in Brussels. He had already stated this ambition in a letter of 1913:

> "In a bit I plan a visit to Brussels to study some of the sites I have long admired: Ste. Gudule, La Bourse, Le Palais de Justice—three different autocrats, but autocrats nonetheless—deserving a bath in the acid! What a superb and nearly virgin city for an indiscreet and somewhat artistic eye![95]

De Bruycker paid partial tribute to the cathedral of Ste.-Gudule and the Bourse (the stock exchange) more through implication than description. In his 1925 etching *Le Vieux Bruxelles* (Old Brussels, cat. 30) the cathedral sulks in the distance, personifying the centuries in the face of transient events much like the Castle of the Counts in De Bruycker's pre-war etchings. As De Bosschère put it, in discussing De Bruycker's approach to cities:

> One doesn't have to be an archaeologist to realize that accumulations of buildings begin to live in the course of many years. Then an indescribable connection develops between people and things. This connection consists of thousands of nuances, which can be detected and observed through these feelings.[96]

De Bosschère was correct to detect an intangible quality in the relationship between old cities and their residents in De Bruycker's work. This puts De Bruycker in the otherwise unlikely company of Giovanni Battista Piranesi and Charles Meryon, who also etched ancient cities (Rome and Paris, respectively) whose ruinous age is enlivened by the random activity of their current population.

De Bruycker's busy etching of the vicinity of the stock exchange, *Bruxelles, Jour de Bourse* (Brussels, Stock Exchange Day, cat. 40) does not show the exchange building. While it is difficult to reconstruct, this image of 1930 appears to be a view from a window within the Bourse itself, with Mausstraat to the left, looking across Anspachlaan to Van Praetstraat. If this is correct, then the image has been reversed in the process of etching. Despite his stated intentions, De Bruycker has preferred to turn his back on the enormous neoclassical stock exchange to depict the riot of advertising and the turmoil of the populace at this convergence of six streets. Unlike Ensor's somewhat analogous drawings preparatory to his *Entry of Christ into Brussels*, the texts in this image do not offer access to a personal vision or subversive allegory, rather, they and the frenetic activity around them affirm a tangible pandemonium.[97]

In 1925, the year after his election to the Royal Academy, De Bruycker travelled to Paris to visit his friend Frans Masereel. De Bruycker was somewhat covetous of Masereel's fame, won (to De Bruycker's mind) through Masereel's decision to locate himself near the hub of activity in Paris.[98] Apparently it was Masereel who inspired De Bruycker to take on a series of architectural studies.[99] This took the form of a group of large and complex images of cathedrals. The plates are of the cathedrals of Paris (1926), Antwerp (1929, cat. 38), Rouen (1930, cat. 39), Bourges (1931), and Amiens (1932). These, and some of his other city views, such as *La Porte St. Denis, Paris* (Le Roy 154) offer a more splendid civic face than we are used to in De Bruycker's art, and it is possible that these were in fact done to cultivate patronage outside of Ghent. However, he made no concession to his proclivity to populate his city scenes with masses of common people at street level, and he exercised a critical eye in sizing up these new subjects, stating for example that Rouen is "a curious and almost too picturesque city that, in my opinion, lacks character, structure, and force."[100] These imposing plates seem to have nearly exhausted De Bruycker. By 1930 he complained in a letter of the weary and taut tendons he developed while making his large plate of Bourges.[101]

Perhaps the most successful post-war works by De Bruycker are his intimate scenes of his immediate environs and of his most enduring source of subject matter, the predominantly unchanged features of old Ghent. Many of these etchings bring us into the immediate world of the artist. *L' Echaugette quai St. Pierre, Gand* (The Warming Tower, Quai St. Pierre, Ghent, cat. 31 and 32) of 1925 shows a tower visible from the artist's residence at 8 Sint-Pieterskaai (Quai St. Pierre).[102] In the winter boatmen would warm themselves at the fire in this tower.[103] In the final state of this etching (cat. 32) the artist trimmed the copper plate at the bottom and the right, removing his self-portrait in which the artist's etching hand is given exaggerated importance. It has been proposed that the lovely views from his house and its immediate environs were in part a result of his taking refuge in his home in reaction to his mother's long illness and death in 1928 (cat. 35, 36, 37).[104] It may also be that his 1928 etchings of the Church of Sint-Niklaas (cat. no. 33) were done from earlier drawings during this period of grief, just as his *Mendiants* of 1928 was based on his earlier woodcut illustrations to *Ulenspiegel*.[105]

This period of reassessment approximately coincides with the onset of De Bruycker's scrutiny of himself. From 1925 on he produced a remarkable series of figure studies, including self-portraits and studies of models (clothed and unclothed). Frequently these incorporate the trappings of the studio or flights of imagination (cat. 45, 46). A buddhist statuette, an African mask, boxes of cigarettes, and prints and drawings by De Bruycker can be identified among the items in his studio.[106] The self-portraits that incorporate elements from his imagination typically focus on primary aspects of his inspiration, such as his study of around 1936 showing himself drawing compulsively before the church of Sint-Niklaas as if he and the edifice are somehow inseparable (cat. 45). Only on rare occasions was De Bruycker satisfied to show himself unencumbered with the minutia of his environment, such as the unusually intimate self-portrait sent to the Venice Biënale in 1934 (cat. 41).

In 1932 De Bruycker's first of several portfolios appeared, *Sites et Visions de Gand*, a collection of twenty-two etchings with a preface by Grégoire Le Roy, who would publish a catalogue raisonné of De Bruycker's prints the following year. In the best works in *Sites et Visions* De Bruycker shows that at sixty-two he was still perfecting his skills. Plates such as *Maison Bourgeoise* (Bourgeois House, cat. 42), *L'Église St.-Michel* (The Church of St. Michael, which is next to the Dominican Pand, cat. 43), and *Le Quai de l'Ecluse* (Floodgate Quay, cat. 44) reveal a renewed interest in the prints of Baertsoen, Whistler, and Rembrandt. *L'Église St.-Michel* is one of the few prints by De Bruycker that makes specific reference to political issues. The plate is essentially a re-working of the 1926 etching *Tweeslachtige Stad* (Schizophrenic City, also titled *Petite Ville Nerveuse*, Small Anxious City, fig. 20), which was done in response to De Bruycker's witnessing a demonstration over the language issue that had been brought to a head by the controversial decision to make Ghent University a Flemish Language institution. In *Tweeslachtige Stad* Ulenspiegel reappears in De Bruycker's work, in Chabot's words, to kick out the "franskiljon" (pro-French or Francophone Fleming) factions.[107] Only the words "HAUTES ETUD[ES]," (advanced studies) remain in *L'Église St. Michel* to evoke, however subtly, the language issue; gone is Ulenspiegel and the reference to "Smetse Smee," an old Ghent tale about a blacksmith that was recounted in Charles De Coster's *Légendes flamandes*.[108]

As if consciously reining in his career, De Bruycker issued two more portfolios late in life and planned a third. The first of these was a summing up of his life-long dedication to the church of Sint-Niklaas, *L'Église St.-Nicolas à Gand* with ten original etchings (fig. 21) and 20 reproductions of drawings. This album, which included an introduction by Grégoire Le Roy, occupied De Bruycker from 1936 until 1938, when it was issued. It is significant that De Bruycker dedicated a great deal of effort to describing the Church of Sint-Niklaas, which catered to the lower classes in Ghent, to the virtual exclusion of the cathedral. It may also be significant that the church interiors are nearly empty and do not concern the performance of religious rites.

Fig. 20. Jules De Bruycker, *De Tweeslachtige Stad* (Schizophrenic City), etching, 1926. Museum voor Schone Kunsten, Ghent.

Among the drawings that are reproduced in the album are several staggeringly vertiginous and volumetric renderings of the church interior (cat. 47, the original drawing) in which one can almost hear the echoing of wooden chairs scraping the stone pavement as a few figures move about in the empty hulk.

Despite increasingly difficult health problems (primarily a weakening of his bones) that demanded that De Bruycker dedicate most of his efforts to drawing rather than printmaking, he managed to etch a selection of twenty-two drawings culled from the dozens he had executed on the outdoor terrace of the café Wilson in 1940. Chabot describes this situation:

> Life has become hard. De Bruycker is sick. He is brought by taxi to sit on the steps in front of a café. He nestles into a corner and chain-smokes his cigarettes. Concealed, he draws like one possessed. The other guests sit with their backs toward him, he observes them and penetrates their character.[109]

An Eye on Flanders: The Graphic Art of Jules De Bruycker   35

Fig. 21. Jules De Bruycker, etching from the portfolio *L'Église St. Nicolas Gand*, 1938. Museum voor Schone Kunsten, Ghent.

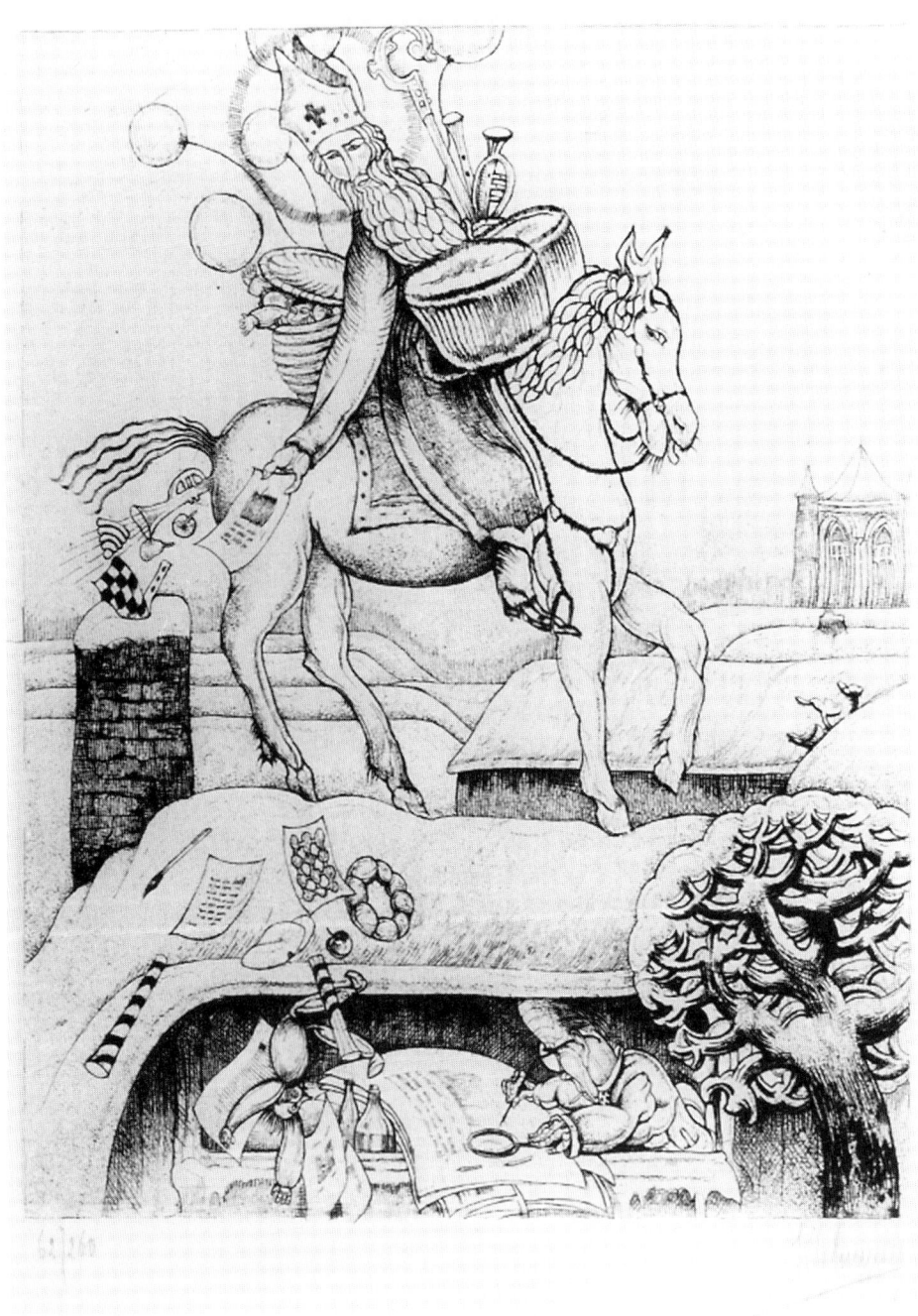

These studies of people from various walks of life, often done from the back, appeared in 1942, as *Gens de chez nous* (People from Around Here, cat. 48).

Shortly after Germany attacked Belgium in May 1940, a bridge near De Bruycker's residence was blown up, damaging his home. Chabot reports that the artist dwelled increasingly on humankind's self-destructive tendencies, and, further, that he feared that his nightmarish visions of the First World War were coming to pass.[110] Chabot also records that De Bruycker would scream out during aerial attacks, "They are going to blow everything to pieces! They are going to destroy everything!"[111] The impact of military occupation on the seventy-year-old artist must have been chilling, especially considering that he had lived out the previous occupation of Belgium vicariously. It is ironic that throughout the years of occupation German print enthusiasts would occasionally arrive in a black Daimler-Benz at De Bruycker's residence, among them the director of the Berlin Museum. On one occasion De Bruycker was even asked to supply a drawing for propaganda purposes by the Germans, but he refused.[112]

In a superb gesture of defiance and perhaps daring, fully in keeping with De Bruycker's lifelong practice of quietly recording his surroundings, he began his final portfolio, which was to be called *Gens pas de chez nous* (People Not from Around Here, cat. 49, 50, 51). This portfolio, issued posthumously in 1975 under the title *oorlogsetsen en -schetsen* (Wartime Etchings and Drawings), includes prints pulled from the six original plates that De Bruycker finished etching before his death on September 5, 1945, as well as reproductions of eight drawings that were made in preparation for the series. These compositions superficially resemble the studies in *Gens de chez nous*, however, a closer examination reveals that the figures enjoying their cocktails, cigarettes, and leisure in the terrace are the occupying forces, not the citizens of Ghent. This is a far more horrifying vision than the Bruegelian landscapes that De Bruycker had invented during the First World War, for it is an eye-witness account that testifies to our strange human ability to participate as individuals in war while remaining blithely unscathed. It is this unflinching integrity to what is seen, tempered with an ironic undertow, that marks much of De Bruycker's career. It is also this quality that van de Woestijne doubtless had already detected in 1912 when he wrote, "De Bruycker may also be called: an Eye—but an eye in which the optic nerve plunges into self-denying discontent, into self-mocking bitterness."[113] Van de Woestijne, who was writing before either World War, may have been mistaken, however, to see De Bruycker's cynicism as self-directed; rather, De Bruycker's cynicism should be understood as an extension of his eye, an extension that looks outward, not inward.

NOTES

1. Baedeker, *Belgium and Holland*, p. 75.

2. The Battle of the Golden Spurs of 1302 is still commemorated annually on July 11 as a holiday for Flemish-speaking Belgium.

3. Van Lerberge, *De Geschiedenis van Bond Moyson*, pp. 11–12.

4. Goddard, *Les XX and the Belgian Avant-Garde*, pp. 22–24.

5. De Bosschère's major article on De Bruycker was published in Flemish, but all of his literary works were composed in French.

6. Bonnel, "Herinneringen aan Juul de Bruycker," p. 1064: "Nooit of nimmer heb ik, in al die vele jaren, politieke of taalkwesties met hem besproken. Wel kon hij, af en toe, schertsend of sarkastisch een schot—vóór of tegen—aflaten. Verder ging het niet."

7. See the over-1500 page catalogue, Ghent, Museum voor Schone Kunsten, *Gent, duizend jaar kunst en cultuur*. See also Baillieul, *Een Stad in Opbouw. Gent voor 1540*, and Dambruyn, *Een Stad in Opbouw. Gent van 1540 tot de Wereldtentoonstelling van 1913*.

8. Mussche, *Gent en zijn etser-teekenaar Jules De Bruycker*, p. 10: "In den wachttoren en in de kapel ronkt en rookt een katoenfabriek, krijschen de machines, zwoegen mannen en vrouwen en kinders in vale killigheid, erger dan beesten. En aan de koude harde steenen van het slot kleven als melaatsche knobbels en zweren honderd erbarmelijke krotten en kroegen: onvergelijkelijk beeld der klasse-tegenstellingen."

9. Frommer, *A Masterpiece Called Belgium*, pp. 182–86.

10. Fris, *Histoire de Gand*, p. 353. This study by the city archivist was first published for the celebration surrounding the 1913 International Exposition in Ghent.

11. Fris, *Histoire de Gand*, pp. 355–56, and see the extensive treatment in Capiteyn, *Gent in Weelde Herboren*, pp. 9–43.

12. I thank John De Bruycker for pointing this drawing out to me.

13. For the Dead City Theme see Friedman, *The Symbolist Dead City*, and Pudles, "Fernand Khnopff, Georges Rodenbach, and Bruges, the Dead City." However unlikely, Hellens claims that he had not read *Bruges-la-Morte*; Hellens, *Documents Secrets*, p. 38.

14. Hellens, *Documents Secrets*, p. 39:
Je me procurai son adresse et me décidai à aller frapper à sa porte. A mon grand désappointement, il habitait une petite maison proprette dans une rue quelconque de la ville neuve, au lieu de cette vieille tour que j'avais remarquée près du canal et où j'aurais voulu qu'il perchât. Sans doute allais-je me trouver en face d'un être mal vêtu, dont le visage terreux et maigre et les longs cheveux défaits indiqueraient tout de suite la qualité d'artiste... Un petit homme d'aspect soigné, proprement habillé et d'un visage tout à fait normal, vint m'ouvrir...
Où avais-je déjà aperçu cette figure? Je me souvins. Mes parents habitaient une maison bourgeoise, entre ville et campagne. Un après-midi, j'étais occupé à bûcher les matières du prochain examen, quand on frappa à la porte. Je répondis avec mauvaise humeur et jetai à peine un coup d'oeil sur l'importun, un homme de trente ans environ, d'extérieur tout à fait quelconque, et dont les traits, au premier abord, semblaient dépourvus de caractère. Il portait une courte barbe soigneusement taillée et tenait en main un chapeau melon.
— Je suis le tapissier, me dit-il d'une voix effacée.

15. Le Roy, *L'Oeuvre gravé de Jules De Bruycker*, pp. 11–12:
Ne suis-je pas moi même un tapissier raté? Les mauvaises leçons sont sans doute celles qu'on écout le mieux. Moi aussi je rêvai de faire des tours et des cabrioles; d'une façon plus modeste, bien etendu, avec mon crayon—car

je n'avais de dispositions ni pour la scène ni pour le cirque.
Je dessinai tant et plus, mon père en profita pour m'envoyer à l'Académie. La science des styles, pensait-il, est utile au tapissier.
Nous étions à l'aube des temps où les bourgeois rêvaient d'une salle à manger Renaissance et d'un Salon Louis XVI.

16. Le Roy, *L'Oeuvre gravé de Jules De Bruycker*, p. 15:
Patatras! Du coup je sentis s'éteindre la petite aureole dont je me croyais paré. Pour ma ville natale, plus de 50 eaux-fortes, de nombreuses aquarelles et de milliers de dessins ne comptaient pas. J'étais toujours *le tapissier*, l'humble tapissier!

17. Karel van de Woestijne, "Jules De Bruycker," 1922 (I), p. 1: "Wij waren immers geene artiesten wij: gij waart een behangersjongen (ik mag het toch verklappen?), ik een studeerende op den eersten trap der universiteit."

18. Chabot, "Jules De Bruycker," 1963, p. 124:
De direkteur van de Akademie voor Schone Kunsten, Delvin, schreef op 12 mei 1916 aan Van Herrewege: "Lang geleden heeft De Bruycker mij eens binnen gelaten—ik weet niet meer hoe of waarom—op de kamer waar hij werkte. Ik was onmiddellijk aangegrepen, totaal. Overal tekeningen op stukjes los papier, onverzorgd en opzettelijk onafgewerkt om de belangstellenden te bekoren. Alles was gebruikt, onverschillig hoe, om toch maar een overdadige en overvloedige gang van gedachten uit te beelden. Daar zag ik de menselijke ellende blootgelegd, zoals ze is, scherp of verdoken.
En ik herinner mij de rustige figuur van De Bruycker. Alleen de vluchtige ogen verraadden zijn onrust. Ik herinner mij zijn vlammende blik, wanneer hij staande hield geen karikatuur te hebben gemaakt.
'Een kunst om te doen lachen? zei hij. Mijn ziel schreit wanneer ik de mensen zo uitbeeld. Ik zie ze zo, en zo zijn ze. Ik heb er medelijden mee, eindeloos medelijden.'"

19. Le Roy, *L'Oeuvre gravé de Jules De Bruycker*, pp. 13, 14:
En 1902 je changerai de quartier et m'installai au *Patershol*—un ancien couvent fort délabré—où plus d'un artiste avait établi sa demeure. Ces habitations de fortune donnaient presque toutes sur une vaste cour intérieure, dispensatrice d'une lumière favorable. C'était un milieu pittoresque, non seulement par la vétusté des bâtiments, mais aussi par la variété des locataires: des artistes et d'incorrigibles pauvres et poivrots. C'est là que je m'attaquai aux aquarelles de grand format.

20. Robert Hoozee, *Veertig Kunstenaars Rond Karel van de Woestijne*.

21. The title of the progressive Flemish journal *Van Nu en Straks* has been translated as "From Now On," see Goddard *Les XX*, pp. 354–55.

22. Karel van de Woestijne, "Jules De Bruycker," 1912, p. 902, based in part on the translation in Boyens, *Flemish Art*, p. 26:
Om dat kachelken heen zaten wij, op nogal kreupele stoelen, en rookten steenen pijpen. Herinner u, Herman Teirlinck, die toen in Gent studeerdet, hoe wij er krakeelden, om dat intieme kachelken, over Shakespeare-kritiek. En gij, alle andere makkers, hoe wij er Kropotkine uitpluisden. En gij, Julius de Praetere-zelf, gij thans Herr Direktor des Zürcher Kunstgewerbe-Museums, herinner hoe wij gloeiden voor eene nieuwe boekdruk-kunst; hoe wij zonder geld en zonder zekerheid, de uitgave "Werk" aangingen, die, na de eerste reeks van "Van Nu en Straks," al bleef zij bij haar eerste nummer, de schoonste, en, naar druk en ornamenteering, zeker het meest logisch-typografisch, het meest bewust-samengestelde en -uitgevoerde tijdschrift was... Herinnert u alleen, hoe we, (al dagteekent uit dien tijd onze enquête, herberg aan herberg, naar het beste Oudenaardsche bier), leefden van een onrustig maar rusteloos, van een aarzelend maar prachtig geestesleven!...
In de zaal stond verder, midden in tafels en schilderezels, de logge hand-pers, waar De

Praetere zijne inzichten aangaande drukkunst verwezenlijkte. Aan lange touwen droogden de vellen van Stijn Streuvels' "Lenteleven," waarvan de eerste uitgave hier uitging.

23. Karel van de Woestijne, "Jules De Bruycker," 1922 (I), pp. 1–2:
Zooals andere kunstenaars in den Pand, het Rolleken of de Vrouwebroers hun intrek hadden genomen, woonde hij in één der lokalen van het Patershol. Pand, Rolleken, Vrouwebroers en Patershol waren verweven kloostercomplexen. De Pand hing vroeger af van de Sinte-Michielsheeren; het Rolleken dankte zijn naam aan de ronde, door een schot gescheiden holte, waarin men (op gezag van niemand minder dan Jean-Jacques Rousseau) de jong-geboren kinderen kwam neêrleggen, waar men in het vader- of moederhart geen plaatsing voor had. De broeders van Onze Lieve Vrouw hadden hun naam gegeven aan het derde reuzengebouw. Het Patershol droeg een naam zonder dubbelzinnigheid; het was het inkt-zwarte hol waar in verleden eeuwen honderden paters woonden. Vooral de twee laatste lokaliteiten, hoe vervallen ook, konden doorgaan voor architectonische merkwaardigheden. Gelegen in de logge schaduw van het haast duizend jaren oude Gravensteen, waar een wriemeling van werklieden op dat tijdperk uitvoerige verstellingswerken aan het doen was; als eindblokken van enge, lange, bochtige stegen waar in het midden een greppel zijne walmende wandeling maakte en waar, uit de oneindig-hooge en overhellende huizen de geur van zeeploog en roôkool als de natuurlijke uitwaseming was van de wijven die men er kijven en de kinderen die men er drenzen hoorde, stonden zij in hunne bonkige onaanroerbaarheid. De Vrouwebroers blonken uit door eene weldoende zindelijkheid. Een ruime poort voerde naar eene binnenplaats die, links, den breed-bladigen wasdom vertoonde van hooge zonnebloemen, op de drempel van kleine huisjes waar het gezin van oude lieden zijn intrek had. Recht, in de klare gang die aan die zijde de binnenplaats en haar grazige keien afsloot, was een breede eikenhouten trap met sierlijk snijwerk, die naar de tot atelier ingerichte cellen leidde. Men vond er doorgaans schilders met eene decoratie; het was een deftige gelegenheid. Het voornaamste gebouw in het Patershol, dat lag vlak over eene verdachte herberg die tot tien uur in den avond vol bange doodschheid was maar die in den nacht aan het huilen ging alsof men er telkens iemand vermoordde onder een onherroepelijk gebod van helsche machten,-het voornaamste gebouw van het Patershol zag zijne binnenplaats ingenomen door een kuiper, die onder zijne dennenhouten vaatjes een spanen vuurtje deed branden dat oogen en keel aanbeet. Links, onder het oude schaliedak dat op een fraaigebeeldhouwde kepering inzonk, draaide een steile steenen trap, aan de hulp van een vet-blinkend touw, waar de schildersateliers op openden. Het waren groote, blauw-gekalkte ruimten, met eenige welsprekende slordigheid ingericht. Het roode sprietoog van een decoratie lonkte daarin door geen enkel knoopsgat. Daarentegen boden onop-gemaakte bedden met bedenkelijk linnengoed eene gastvrije slaapsgelegenheid. Opmerkelijk was hier het gebrek aan schilderijen: arme schilders denken maar zelden aan werken. Maar men zag onafgewasschen borden die de overtuiging wekten dat hier toch nu en dan gegeten werd. De wanden deden voor die van Italiaansche paleizen niet onder: menig naakt figuur was er, in een oogenblik van grootheidswaanzin, met houtskool op nagebootst. Een stapeltje boeken lag er onder een verroest revolver: onaantastbaarheid van de vruchten des geestes. En verder was er de ellende van een klagend man.

Ik ben maar met het leelijkste begonnen. Daaronder was onder meer de cel van iemand, die het zoowaar tot professor aan de teeken-academie had gebracht. Hij had een teringlijdende vrouw op een kamer in stad : daardoor had hij hier geen bed noodig. Hij had het vervangen door een harmonium, dat de rol speelde van prikkel. Schiller omringde zich met rotte appelen, Puvis de Chavannes streek zijn beenen in met eau de cologne, Beethoven goot water over zijne polsen; deze speelde op een valsch klinkend harmonium het Ave Maria van Gounod. Hij had dat noodig om kruiswegen te schilderen, want dat was zijne aanhoudende bezigheid. Nauwelijks was eene graflegging af, of er werd een doek opgespannen voor eene Veroordeeling door Pontius Pilatus. Met de grootste regelmatigheid werden daar tusschenin de twaalf andere statiën geschilderd. Nooit kwam in de schilderijen de minste wijziging, noch aan teekening, noch aan kleur. Wij wisten heel goed wanneer hij aan een smartelijken blik-toetssteen van zijn kunde-zou beginnen uit den top van een rond kwastje: hij speelde een Ave Maria, en de inspiratie was er. Wij hadden met hem een soort minachtend medelijden: zijne vrouw had tering, zijne vaalbleeke kinderen bij aanhouding een snotneus. Het was onder de inspiratieve tranerigheid van het Gounod-harmonium dat ik voor het eerst mijne hand legde in die van Jules de Bruycker, behangersgast. Het was in de laagste der zalen; hooge ramen lieten den bitteren geur door van het spanen vuurtje dat den buik verwarmde van de dennen tonnetjes, waar men appelen in opzenden zou naar Engeland, hetgeen te dien tijd eene Gentsche bezigheid was die ik nooit recht heb begrepen. In deze ruime kamer met rijzige gewelven, waar men de vestigia van een altaar vermoedde, was zoo goed als niets dat aan schilderkunst denken deed. Maar men vond er eene Gotische grammatica, de Engelsche oefeningen van Stoffel en eene kritische uitgave van *Kabale und Liebe*: mijn eigendom. Gemeenschappelijk eigendom was een stapel nummers van *Les Temps nouveaux* en van den *Mercure de France*. Er was ook een zeer fraaie pijp die ik aan de gemeenschap cadeau had gedaan, en die geen ander pleizier had dan te soppen. Dit lokaal werd te dien tijde vereerd met het bezoek van Stijn Streuvels en Herman Teirinck. Te dezer gelegenheid werd de roode tegel-vloer schoongeveegd door een daartoe uitgenoodigd model. Zij was een dutsig meisje vol gedweeheid. Zij dacht dat men haar ontbood om to poseeren: het was alleen om to vegen. Dat zij zich niet hoefde uit te kleeden was voor haar als eene teleurstelling. Nooit heeft zij voor het vegen een cent willen aanvaarden...

In deze zaal en in deze omstandigheden ontmoete ik voor het eerst Jules de Bruycker. Hij had het niet prettig: een kleine tien jaar ouder dan ik, begaafd met een talent dat toen reeds al zijne vrienden waardeerden, al had hij nog nooit eene etsnaald in hand gehad- alleen behangersnaalden placht hij te hanteren-sprong hij bij ons binnen, den riem van een ronden, rooden, tapijtenzak gesneden in den schouder. Hij moest zorgen voor moeder en zuster: hij werkte met gespannenheid en bitterheid. Hij was een schuchtere en een scepticus: hij lachte met schamperheid, als men hem zei dat hij meer geld zou hebben kunnen verdienen als teekenaar dan als tapissier. Hij was oud genoeg om aan geen hersenschimmen toe te geven. Er was trouwens iets dat hij maar al te goed wist: een goed schilder zou hij nooit zijn geworden; hij was zoo goed als kleurenblind. Hij, die toen reeds een grandioos caricaturist was, met al de eigenschappen van een Daumier; van Daumier miste hij de broeiende kleur. -Soms bracht hij ons teekeningen ; een enkel maal verzocht hij zelfs onzen vriend één zijner teekeningen te aquarelleeren (het was, geloof ik, zijne eerste bestelling). Toen zagen wij hem aan: zijn clowneske kop verwrong en trilde om den mond. Zijne wonderlijke puntneus die waarlijk *kijkt*; zijne naar binnen geboorde oogjes; zijn smalle en wonder beweegbare lippen; en heel dat schrale en soepele lichaam: zij waren ééne negatie; dat alles beteekende: niets, niets.

En nochtans kon hij niet buiten teekenen. Meer dan eene obsessie, was het hem eene behoefte. Ontmoete hij op straat een type dat zijne aandacht trok, hij achtervolgde hem, een papiertje in de handpalm, het potlood tusschen de vingeren. Hij vergat het behangerswerk: hij was de bezetene der kunst.

24. Chabot "Jules De Bruycker," 1963, p. 125.

25. Karel van de Woestijne, "Jules De Bruycker," 1912, p. 903.

26. Eeckhout, *Jules De Bruycker*, p. 7.

27. Hellens, "La Cuisine de fous," in *Les Hors-le-Vent*, pp. 42–43. The passage is also given in Flemish translation in Chabot, "Jules De

Bruycker," 1963, pp.125–126 (Chabot discusses Emile Van Vooren on p. 124):
Le dessinateur occupait dans la cuisine du concierge une place marquée. Monsieur Charles l'admirait avec une naïveté d'enfant devant un prêtre, bien que le privilégié ne parût pas s'en soucier, se contentant de trouver le feu excellent, le cigares fins, la place commode et lucrative. Son nez mobile de fourmilier semblait sans cesse chercher le fumet sensuel des choses. L'atmosphère surchauffée de cette cuisine, où passaient des effluves d'animalité dans le relent gras des nourritures, convenait à ses instincts flaireurs; il y faisait ses meilleures études, tout en ayant l'air uniquement de se chauffer et de causer. Les murs s'écaillaient de croquis inachevés, de charges folles où ni la bonhommie du concierge ni le gâtisme de sa chienne n'étaient respectés. Derrière la loge, dans un débarras, des cartons bourrés d'études s'empilaient; le concierge se chargeait de la vente. Les collectionneurs venaient de loin le trouver! Et Monsieur Charles, très fier, disait: "Je travaille pour mes artistes!"
See also Bonnel, "Herinneringen aan Juul de Bruycker," pp. 1051–54 for this locale.

28. For Baertsoen see Eeckhout, *Albaert Baertsoen*.

29. Le Roy, *L'Oeuvre gravé de Jules De Bruycker*, p. 13:
Un jour, ma flânerie me conduit au musée où je découvre une planche de Baertsoen. Ce fut une révélation! Mon emballement pour de procédé fut soudain, irrésistable. Je me lançai chez *Fritz Van Loo* que je savais initié en cet art. Il me donna ma première leçon.
Son atelier était au *Pand*, rue de Valée. Encore un couvent désaffecté.

30. For Van Loo see *Dictionnaire biographique* p. 380.

31. Chabot, "Jules De Bruycker," 1963, p. 115, who gives De Bruycker's age as 30, not 35 when he started to etch. For Van Campenhout and Maison Dietrich see Goddard, *Les XX and the Belgian Avant-Garde*, pp. 80, 87.

32. I want to thank Robert Hoozee for taking the time guide me through Ghent in order to identify the sites.

33. In 1992, the *Veergrepe* (or *Veergreep*, as it appears on today's maps) was partially levelled and turned into a public park.

34. Goddard, "Print Culture in Nineteenth-Century Belgium," in *Les XX and the Belgian Avant-Garde*, pp. 75–97, p. 90.

35. Karel van de Woestijne, "Jules De Bruycker," 1912, pp. 912–15.

36. For the episode of van de Woestijne and de Sadeleer, see Gustave van de Woestijne, *Karel en Ik*, pp. 83–84.

37. Mussche, *Gent en zijn etser-teekenaar Jules De Bruycker*, p. 17.

38. Bonnel, "Herinneringen aan Juul de Bruycker," p. 1052.

39. Decavele, *De Opera van Gent*, pp. 141, 188–89, for photographs of the Opera Paradise (before and after restoration completed in 1993).

40. This volume of loose mounted drawings has been titled *Schetsen "opstrate" I* (*Sketches from the Street I*). Two of them are dated 1914.

41. For de Praetere's father see Gustave van de Woestijne *karel en Ik*, p. 34. In *Vieux marché à Gand* the artist shows a view of Sint-Jacobs from the West but has substituted the tympanum from the North portal; and in *Vieux marché en Flandre* Sint-Jacobs is seen from the Vlas markt to the south, with the view modified so that the Baudeloo Chapel appears to the upper left.

42. Karel van de Woestijne, "Jules De Bruycker," 1912, pp. 910–11: Lange en geerige vingers wegen vreemde sleutels, tillen naar boven de ijzeren roosters waar roest verteert cierlijke arabesken, betasten Damascus-poken of, eenvoudiger, een paar niet al te zeer

aangevreten strijk-ijzers, tenzij het een zeldzaam stel achttiende-eeuwsche romans was of grauw-omkafte pamfletten uit de zestiende, de reeks der volledige werken van Buffon of eene verzameling mode-platen; of nog: eene heilige relikwie, eene bed-pan, een keizersportret, een pannen-lap tot het poetsen der kachels, een schutterij-uitrusting, een lamp een revolver, een middeleeuwsch stads-register, oud porselein, afgedragen kleederen (naar dewelke groote vraag is), muziek-instrumenten of een koopje voorhistorisch schoen-smeer.

43. Karel van de Woestijne, "Jules De Bruycker," 1912, p. 935.

44. *Bouwen door de eeuwen heen. 4na. Stad Gent* vol. 1, pp. 271–72, and see the color plate.

45. Guislain, *Le palais de Justice*.

46. Guislain, "Jules De Bruycker aquafortiste et poète," p. 1:
—J'ai travaillé la bas. Pas assez. J'aurrais aimé dessiner le Palais de Justice. Vous l'avez appelé le Mammouth? Les architectures étagées du Mammouth m'ont toujours tenté. Il est magnifique, ce temple immense ruisselant de lumière. Cette superposition de formes, cet étagement de pierres et de terrasses, quelle splendeur! Quelle chose à faire, comme nous disons. J'y suis allé, j'ai cherché. De braves gens de la rue Haute et de la rue des Minimes m'ont offert l'hospitalité de leur grenier et de leur mansardes. Mais la chance ne m'a pas favorisé. Un obstacle est toujours venu s'interposer entre votre Palais de Justice et moi. Regrettable! Et la salle des Pas-Perdus! Quelle eau-forte! Quel dessin! Ma femme et moi-même avons pénétré, un jour, par hassard; dans une salle d'audience. Une affaire se plaidait. Une ménagère discutait la note de son boucher. Elle prétendait qu'on lui avait fourni trop d'os et de graisse et pas assez de viande. C'est drôle, la vie. Et dans cette basilique formidable, cette toute petite comédie. C'était humain, trop humain...

47. I thank John De Bruycker for pointing this out.

48. Capiteyn, *Gent in Weelde Herboren*, p. 67.

49. Chabot, "Jules De Bruycker," 1963, p. 133: Personne sauf vous et M. Fornier n'ont vu jusqu'ici le dessin Montage du Dragon. La gravure premier état est presque terminée et je compte sur un résultat supérieur à celui de la planche du château des comtes. Excusez ma vanité! Je gratte comme un enragé, sans relâche, depuis deux semaines. Et je puis avoir cet espoir! Maintenant j'en suis à la morsure. Je vais procéder tout autrement, lentement, une morsure de 6 à 12 heures. Cette eau forte doit être moins matérielle, plus *vision* que le château des comtes. Je le répète, j'ai beaucoup d'espoir et surtout beaucoup de ténacité ou entêtement et quand je faiblis, je descends à la cave et je débouche une de ces bouteilles ... pas d'acide nitrique!

50. Chabot, "Jules De Bruycker," 1963, pp. 132–34. Chabot lists the works exhibited at the biënnale as: *Marktdag te Gent*, *Het Huis Palfijn*, *Het Planten van de Draak*, *Schets*, *De Man van het Belfort*, *De Markt*, *De St.-Jacobsmarkt*, *Graslei (fruitmarkt)*, *Burcht te Brugge (eerste proef)*, *Rond het Gravensteen*, and *De Prondelvrouw*.

51. Buschmann, "Belgian Artists in England," pp. 186–209.

52. Eeckhout, "Bruycker, Jules-François De," pp. 132–33. Not much has been set down about De Leyn. Eeckhout says only, "Sa sensibilité, sa culture, son éclectisme tout autant que son esprit critique exercent sur l'artiste une influence prépondérante tout au long de sa carrière."

53. Chabot, "Jules De Bruycker," 1963, p. 136. Young, *The Paintings of James McNeill Whistler*, text vol., p. lxviii.

54. Chabot, "Jules De Bruycker," 1963, p. 136: Ici on était obligé de rester au studio—défense de travailler en rue. Si on regardait avec trop d'insistance un pont, statue ou

bâtiment, vous étiez suspect! Avec cela le charmant climat, brouillard, pluies et distances formidables. Ce n'était pas facile."

55. De Bruycker's two scenes of Piccadilly Circus were done from the windows of the Swan and Edgar store, not from the street, see Bonnel, "Herinneringen aan Juul de Bruycker," p. 1058.

56. Karel van de Woestijne, "Jules De Bruycker," 1922 (III), p. 1:
Hij leeft thans den meesten tijd in eene hallucinatorische atmospheer: de afstand zet de gebeurtenissen uit, verleent haar eene synthetisch-symbolische beteekenis die wij, in het land gebleven, niet steeds begrepen.

57. Chabot, "Jules De Bruycker," 1963, p. 138, from a letter of November 23, 1918:
J'ai assez bien travaillé. J'ai, sans avoir entendu un coup de fusil, fait des dessins se rapportant à la guerre. Des oeuvres étranges et qui m'ont ouvert pour l'avenir un autre horizon.

58. See, for example, Shapiro, *Painters and Politics*, chapter 6 "Artists at War" pp. 132–72, and Cork, *A Bitter Truth: Avant-garde Art and the Great War*. See also London, Imperial War Museum, *A Concise Catalogue*.

59. Shapiro, *Painters and Politics*, p. 158.

60. This list was compiled in part from Shapiro, *Painters and Politics*, p. 132.

61. Cork, *A Bitter Truth: Avant-garde Art and the Great War*, p. 9.

62. A study by Brendan Burny discussing the discovery of an important source for the skeleton in De Bruycker's print is anticipated in the 1995 Museum voor Schone Kunsten, Ghent, catalogue on De Bruycker.

63. For Ulenspiegel (or Tijl Ulenspiegel, Uilenspiegel, or Eulenspeigel), see Simons, "Tijl Uilenspiegel." For De Coster see De Seyn, *Dictionnaire des Ecrivains Belges*, pp. 380–82; and Klinkenberg, *Charles De Coster*.

64. Anon, "Belgische Kunst te Rotterdam":
Deze ets heet De Dood over Vlaanderenland, en herinnert aan Uilenspiegel de beschrijving-in-woorden van 't vroeger geweld. Hoevele zinnen uit dit boek—óók die van het onsterfelijk land—zijn toepasselijk wéér op den tijd! En hoe heeft deze kunst den ouden geest, den rijkdom het karakteristieke, dat een innig vlaamsche roem is.

65. Willis, *Flemish Renaissance Revival*.

66. Freidman, "Belgian Symbolism," pp. 127–28.

67. Friedman, "Belgian Symbolism," p. 238.

68. Karel van de Woestijne, "De Vlaamsche Primitieven, hoe ze waren te Brugge." His copy of the exhibition catalogue is full of notations, see Somers, *Karel Van de Woestijne* 48, no. 140. For "Ruysbroeck mysticism" see van de Woestijne, "Jules De Bruycker," 1922 (II), p.2.

69. Boyens, *Flemish Art*, p. 41. For the significance of Ruusbroec for Dutch artists at the turn of the century, see Polak, *Het Fin-De-Siècle*, pp. 4, 141.

70. Streuvels, *In oorlogstijd* 92, 115 (Bruegel); 367, 391, 447, 547 (Ulenspiegel).

71. Chabot, "Jules De Bruycker," 1963, p. 161.

72. For Christus see Gustave van de Woestijne, *Karel en ik* pp. 51, 92; and for Mme Minne see the same work, p. 67.

73. For Karel's writing area see Gustave van de Woestijne, *Karel en ik*, pp. 31 and 51. The former described Karel's study in Ghent, which had, in addition to the items listed in Sint-Martens-Latem, a portrait of Max Elskamp and a photograph of the students at the athenaeum. Gustave makes fun of his brother's natty posture in this photograph. For Willem Kloos see *Grote Winkler Prins Encyclopedie*, vol. 13, p. 199.

74. Berg, "Bos(s)chère, Jean de"; de Bosschère, *Portraits d'Amis*; Estang, *Jean de Boschère l'admirable*; Putnam, *The World of Jean de Bosschère*; Warmoes, "Boschère;" Warmoes, *Jean de Boschère*.

75. De Bosschère, *Metsys*; de Bosschère, *La Sculpture Anversoise*, de Bosschère, "Bruegel le drôle." Later in life he dedicated a study to Bosch, published posthumously in 1968, de Bosschère, *Bosch*.

76. De Bosschère, "Jules De Bruycker, Teekenaar en Etser"; "Exhibition by Belgian Artists, the Dowdeswell Galeries, Juni 1917;" "Exhibition of Belgian Art for the Benefit of the Croydon General Hospital."

77. De Bosschère, "Jules De Bruycker, Teekenaar en Etser," p. 1.

78. De Bosschère, *Christmas Tales of Flanders*; de Bosschère, *Beasts and Men*. Although Warmoes, *Jean de Boschère*, p. 55, suggests that these books were only done at the suggestion of Edmund Dulac in order to find some financial success, the choice of subject matter cannot be disregarded.

79. Warmoes, *Jean de Boschère*, p. 47, and Fletcher, "Drawings of Jean de Bosschère," pp. 195, 198–99 (also given in Warmoes, *Jean de Boschère*, p. 57).

80. Archives de l'art contemporain en Belgique (Brussels), no. 40.190, letter to Heer De Graaff, not dated, "de aquarellen die daar te zien zijn buitengewoon interessant. Vlaamsche spreuken en vertellingen op zo 'n mooie wijse geïnterpreteerd!" The exhibition was in the Leicester Gallery, London, Oct. 27, 1915. I thank Micheline Colin for identifying the venue of this exhibition.

81. Bonnel, "Herinneringen aan Juul de Bruycker," pp. 1054–63.

82. From a letter of April 24, 1915, cited in Buenger "Max Beckmann in the First World War," p. 250; citing Rudolf Binding, *A Fatalist at War*, London, 1929, p. 64.

83. For Goulding see Hardie, *Frederick Goulding*, see p. 15 for Goulding Ltd. Goulding died in 1909.

84. Chabot, "Jules De Bruycker," 1963, p. 138; and Eeckhout, "Bruycker, Jules-François De," p. 134.

85. These accomplishments are summarized in Chabot, "Jules De Bruycker," 1963, p. 148. See Vanzype, "Notice sur Jules De Bruycker," for an appreciation of De Bruycker by a fellow academician. There are many reports of an exhibition of De Bruycker's work at the Carnegie Institute, but this may be a mistaken report of the Chicago Art Institute exhibition, see McC., "De Bruycker, Etcher of Ghent, Introduced to America," and anon., "Horrors of War Theme of Many De Bruycker Etchings."

86. Chabot, "Jules De Bruycker," 1963, p. 143; Le Roy, *L'Oeuvre gravé de Jules De Bruycker*, p. 15. For the Pand setting in De Bruycker's works see Simons, *Het Pand*, p. 113.

87. See Simons, *Het Pand*, and De Smet, "Nos Artistes. Mme Cécile Cauterman."

88. For market scenes see Le Roy, *L'Oeuvre gravé de Jules De Bruycker*, numbers 82, 84, 85.

89. The title page clearly announces these illustrations as woodcuts ("bois originaux par J. De Bruycker") but it is hard to imagine that the artist would have mastered the art of wood engraving for this commission and never dabbled in it again. In all likelihood De Bruycker produced drawings that were in turn carved by a wood engraving specialist. Several of the images show the fine white lines characteristic of wood engraving.

90. Chabot, "Jules De Bruycker," p. 148.

91. Hellens, "La Cuisine des fous," in his *Les Hors-le-vent*, pp. 47, 49, 50, 52.

92. See Simons, "Tijl Uilenspiegel;" and *Le Folklore dans l'oeuvre de Charles De Coster*.

Among the Belgian artists who furnished illustrations for De Coster's text are: Félicien Rops (1867, Rops also founded a journal *Uylenspiegel* in 1856), Paul-Auguste Masui-Castricque (1917), Albert Delstanche (1918), Frans Masereel (1926), Joris Minne (1936), and Josef Cantré (date uncertain).

93. Strictly speaking the instruments traditionally associated with lust are the droning instruments, the bagpipes and the hurdy-gurdy; but the more recent accordion shares the qualities of these droning instruments. See Winternitz, "Bagpipes and Hurdy-Gurdies."

94. Marijnissen, "Artiest chargeerde te veel of te weinig."

95. Chabot, "Jules De Bruycker," 1963, p. 150: "Je compte fair sous peu un séjour à Bruxelles pour y étudier plusieurs sites depuis longtemps admirés: Ste Gudule, La Bourse, Le Palais de Justice—trois autocrates différents mais autocrates—valant bien un bain d'acide! Quelle ville superbe et presque vierge pour un oeil indiscret et un peu artiste!"

96. De Bosschère, "Jules De Bruycker," p. 2: Men moet geen oudheidkundige zijn om to beseffen, dat opeenhoopingen van gebouwen pas na lange jaren beginnen to leven. Dan is een niet te omschrijven gemeenschap onstaan tusschen menschen en dingen. Die gemeenschap bestaat uit duizend schakeeringen, welke door deze gevoelens waargeenomen en genoten kunnen worden.

97. For Ensor's drawings see Hoozee, *James Ensor, Dessins et Estampes*, pp. 107–119; see especially Ensor's drawing "LA VIVE ET RAYONNANTE. L'ENTRÉE A JÉRUSALEM," of 1885, Museum voor Schone Kunsten, Gent.

98. Chabot, "Jules De Bruycker," 1963, p. 102, indicates that De Bruycker confided in him his feeling that "Si j'avais fait pour Paris ce que j'ai fait pour Gand, je serais célèbre," and further, that this sentiment returned whenever De Bruycker received a new novel illustrated by Masereel. See also Chabot "Jules De Bruycker," 1963, p. 98.

99. Eeckhout, *Jules De Bruycker*, p. 4 See also Sint-Niklaas, Stedelijk Museum Sint-Niklaas, *Jules De Bruycker*, p. 7, where it is suggested that it was the artist's wife who inspired De Bruycker to travel to France.

100. Chabot "Jules De Bruycker," 1963, p. 157: Ville curieuse mais un peu trop pittoresque à mon avis, manquant de caractère, de structure, de force.

101. Chabot, "Jules De Bruycker," 1963, p. 156: "Après j'ai repris d'autres travaux et les nerfs qui ont été trop tendus sont en ce moment malades et fatigués."

102. De Bruycker lived at Sint-pieterskaai no. 5 beginning in 1908, and Sint-pieterskaai no. 8 from Dec. 1923. A full roster of the artist's addresses in Ghent is given in Chabot, "Jules De Bruycker," 1963, p. 118.

103. I thank John De Bruycker for this information.

104. John De Bruycker, "Biografie," in Sint-Niklaas, *Jules De Bruycker*, p. 7.

105. John De Bruycker, "Biografie," in Sint-Niklaas, *Jules De Bruycker*, p. 7, for the Church of Sint-Niklaaas.

106. See Mussche, *Gent en zijn etser-teekenaar* plates 11–12; Eeckhout, *Jules De Bruycker*, nos. 92–94, 104, 109–110, 113, 128, 148–150; Sint-Niklaas, Stedelijk Museum Sint-Niklaas, *Jules De Bruycker*, no. 75; and Le Roy, *L'Oeuvre gravé de Jules De Bruycker*, nos. 126, 146, 147, 196. I thank John De Bruycker for pointing out instances of De Bruycker's prints and drawings within these studio renderings; apparently the small prints that look like folk prints are proofs for his portfolio *L'Église St. Nicolas à Gand*.

107. Chabot, "Jules De Bruycker," 1963, p. 152.

108. *Le Folklore dans l'oeuvre de Charles De Coster,* p. 73.

109. Chabot, "Jules De Bruycker," 1963, p. 167: Het leven is wel zeer moeilijk geworden. De Bruycker is ziek. Hij laat zich per taxi afhalen en gaat zitten op de stoep vóór een café. Hij nestelt zich in een hoek en rookt, in ketting, zijn sigaretten. Verdoken tekent hij, als een bezetene. De andere gasten zitten met hun rug naar hem toegekeerd; hij observeert ze en dringt diep door in hun persoonlijkheid.

110. Chabot, "Jules De Bruycker," 1963, p.166.

111. Chabot, "Jules De Bruycker," 1963, p.166: "Het is gebeurd dat hij bij luchtbombardementen het uitschreeuwde: 'Ze gaan alles stuksmijten! Ze gaan alles vernietigen!'"

112. Bonnel, "Herinneringen aan Juul de Bruycker," pp. 1063–64.

113. Karel van de Woestijne, "Jules De Bruycker," 1912 p. 920: "Ook De Bruycker mocht worden genoemd : een Oog. Maar een oog, waarvan de zenuwen-bundel duikt in misnoegdheden die zich-zelf verloochenen, in verbittering die zich-zelf bespot."

CATALOGUE

This catalogue is chronological, and follows Le Roy's catalogue raisonné whenever possible. Measurements are in millimeters; for books the page size is given, for drawings the size of the image, and for etchings the size of the plate (plate mark). For works not inscribed with a title the conventional title as given in Le Roy's catalogue raisonné is enclosed in square brackets, as are all translations.

1

[*Confrère*, Colleague], 1906

Etching and aquatint
Inscribed: 3me état
316 x 237

Museum voor Schone Kunsten, Ghent
1906 W
Le Roy no. 3

2
[*Ruelle (Gand)*, Alley (Ghent)], 1906

Etching and aquatint
Inscribed: 4me état
374 x 172

Museum voor Schone Kunsten, Ghent
1906 X
Le Roy no. 6

3

Jour de Marché à Gand [Market Day in Ghent], 1906

Etching and aquatint
Inscribed: aan den kunstvriend Edgar Bytebier
332 x 425

Museum voor Schone Kunsten, Ghent
1978 L
Le Roy no. 9

4

Ruelle (Patershol) [Alley (Patershol)], 1906

Etching and aquatint
376 x 22

J. De Bruycker Collection
Le Roy no. 10

5

[*Veergrepe*, a place name meaning "ferry landing"], 1906

Etching and aquatint
Inscribed (not in De Bruycker's hand): cour flamande
386 x 354

Spencer Museum of Art: Letha Churchill Walker Memorial Art Fund
93.325
Le Roy no. 13

6

En ville morte [In the Dead City], 1906

Book by Franz Hellens; cover and 11 illustrations by Jules De Bruycker; published in Brussels by
G. Van Oest & Co.
260 x 185

J. De Bruycker Collection

7

Three drawings reproduced in *En ville morte* (cat. 6), ca. 1906

Ink wash, graphite, and conté crayon
222 x 79, 201 x 96, 207 x 88

Museum voor Schone Kunsten, Ghent
1932–ET 1–3

8

Le paradis (théâtre) [Paradise Theater], 1907

Etching, soft ground, and aquatint
396 x 286

J. De Bruycker Collection
Le Roy no. 21

9
L'Uilenkot [Owl's Roost], 1907

Etching and aquatint
350 x 250

Museum voor Schone Kunsten, Ghent
1932 X
Le Roy no. 22

10
Vieux marché à Gand,
[Old Market in Ghent], 1907
Le Roy gives the title as
Vieux marché en Flandre

Etching and aquatint
399 x 569

Bibliothèque royale Albert 1er,
Cabinet des Estampes, Brussels
Pl. SII 131543
Le Roy no. 24

11

Théâtre (Lohengrin)
[Theatre (*Lohengrin*)], 1907

Etching and aquatint
298 x 221

Museum voor Schone Kunsten, Ghent
1932 AF
Le Roy no. 25

12
Marchande de bric-à-brac
[Rummage Merchant], ca. 1907
Study for Le Roy no. 26

Conté crayon, graphite,
and gouache heightening
505 x 365

Bibliothèque royale Albert 1er,
Cabinet des Estampes, Brussels
SIV 39443

13
Vieux marché en Flandre
[Old Market in Flanders], 1907
Le Roy gives title as *Vieux marché à Gand*

Etching and aquatint
Inscribed: épreuve unique 2me état
460 x 597

Museum voor Schone Kunsten, Ghent
1932 ED
Le Roy no. 27

14

De Rolweg te Brugge
[The Rolweg in Bruges], 1907

Etching and aquatint with drawing in
ink and watercolor
Inscribed: aan Jean de Moerloose/ het
begin onze carrière
300 x 350

Museum voor Schone Kunsten, Ghent
1965 K
Le Roy no. 29

15
La Maison Jean Palfijn Gand
[The House of Jean Palfijn, Ghent],
1912

Etching and aquatint
605 x 488

J. De Bruycker Collection
Le Roy no. 35

16

Rond het s'Graven Kasteel te Gent
[Around the Castle of the Counts in Ghent], 1913

Etching and aquatint
736 x 625

Bibliothèque royale Albert 1er,
Cabinet des Estampes, Brussels
Max. SIII 21977
Le Roy no. 38

17
De man van t'belfort
[The Man from the Belfry], 1914

Etching and aquatint
247 x 163

Bibliothèque royale Albert 1er,
Cabinet des Estampes, Brussels
Fol. SIII 24303
Le Roy no. 43

18

La montage du Dragon sur le beffroi de Gand (Belgique) [Placing the Dragon atop the Ghent Belfry (Belgium)], 1914

Etching
785 x 610
Inscribed: epr. 1er état/ Hartelijk in aandenken van 1914–1919 in England aan mevrouw en aan den Heer de Graaff

J. De Bruycker Collection
Le Roy no. 44

19

Weer klepte de Dood over Vlaanderenland [Again Death Tolled over Flanders], 1916

Etching and aquatint
785 x 663

J. De Bruycker Collection
Le Roy no. 52

20
Kultur! [Civilization!], 1916

Etching and aquatint
641 x 731

Museum voor Schone Kunsten, Ghent
1921 L
Le Roy no. 53

21
The Harvest [*La Moisson*], 1916

Etching and aquatint
530 x 428

Museum voor Schone Kunsten, Ghent
1921 J
Le Roy no. 55

22

La Tranchée [The Trench], 1916

Etching
641 x 540

Museum voor Schone Kunsten, Ghent
1932 DL
Le Roy no. 56

23
Piccadilly Londres [Piccadilly, London], 1916

Etching
486 x 322

J. De Bruycker Collection
Le Roy no. 62

24

De Slechte Maere [The Grim Reaper], 1917

Etching and aquatint
Inscriptions: épreuve 1r état 1916/Ieperen de slechte maere 1914–1918 1939–1940/affecteusement à ma chére femme/Londres 1916/ épreuve tiré par Goulding Londres
504 x 388

Spencer Museum of Art: Letha Churchill Walker Memorial Art Fund
93.319
Le Roy no. 66

25

Jacobus Alijn (le poilu)
[Jacobus Alijn (the shaggy)], 1921

Etching and aquatint
363 x 237

Bibliothèque royale Albert 1er,
Cabinet des Estampes, Brussels
Pl. SIII 80289
Le Roy no. 91

26

La Légende et les aventures héroiques, joyeuses et glorieuses d'Ulenspiegel et de Lamme Goedzak au pays de Flandres et ailleurs [The Legend and Heroic, Joyous, and Glorious Adventures of Ulenspiegel and Lamme Goedzak in Flanders and Elsewhere], 1922

Book by Charles De Coster, illustrated by De Bruycker; published,
Antwerp: Éditions du Dauphin,
Brussels: Éditions Robert Sand,
Paris: Éditions G. Crès & Cie
195 x 135
26 a: Vol. I p. xi, headpiece for preface, 64 x 81 (image)
26 b: Vol. II frontispiece, 113 x 84 (image)

J. De Bruycker Collection

27

[*Jacobus Alijn* or *De bedelaar*, Jacobus Alijn, or The Beggar], 1921

Graphite and conté crayon
370 x 234

Museum voor Schone Kunsten, Ghent
1932 EW

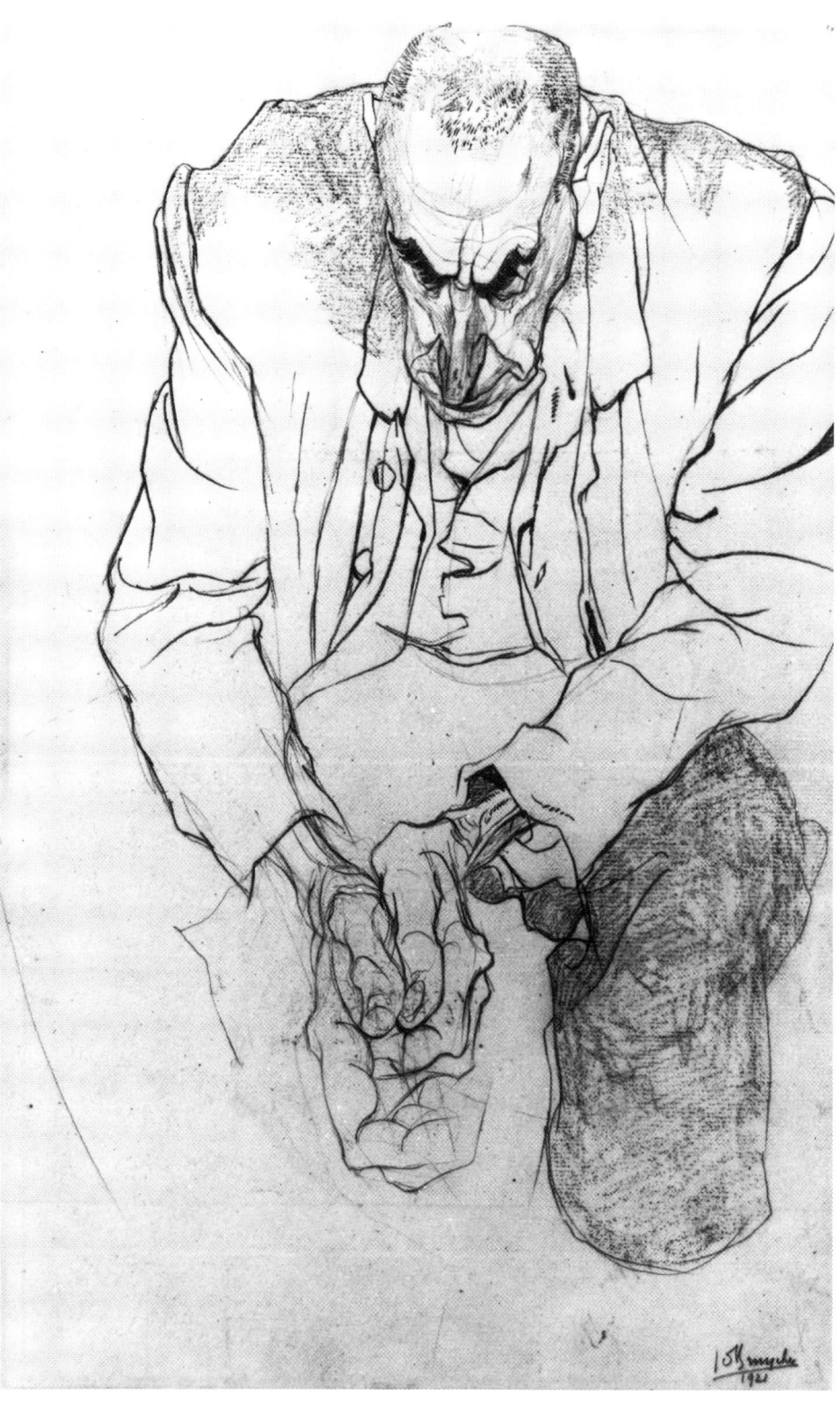

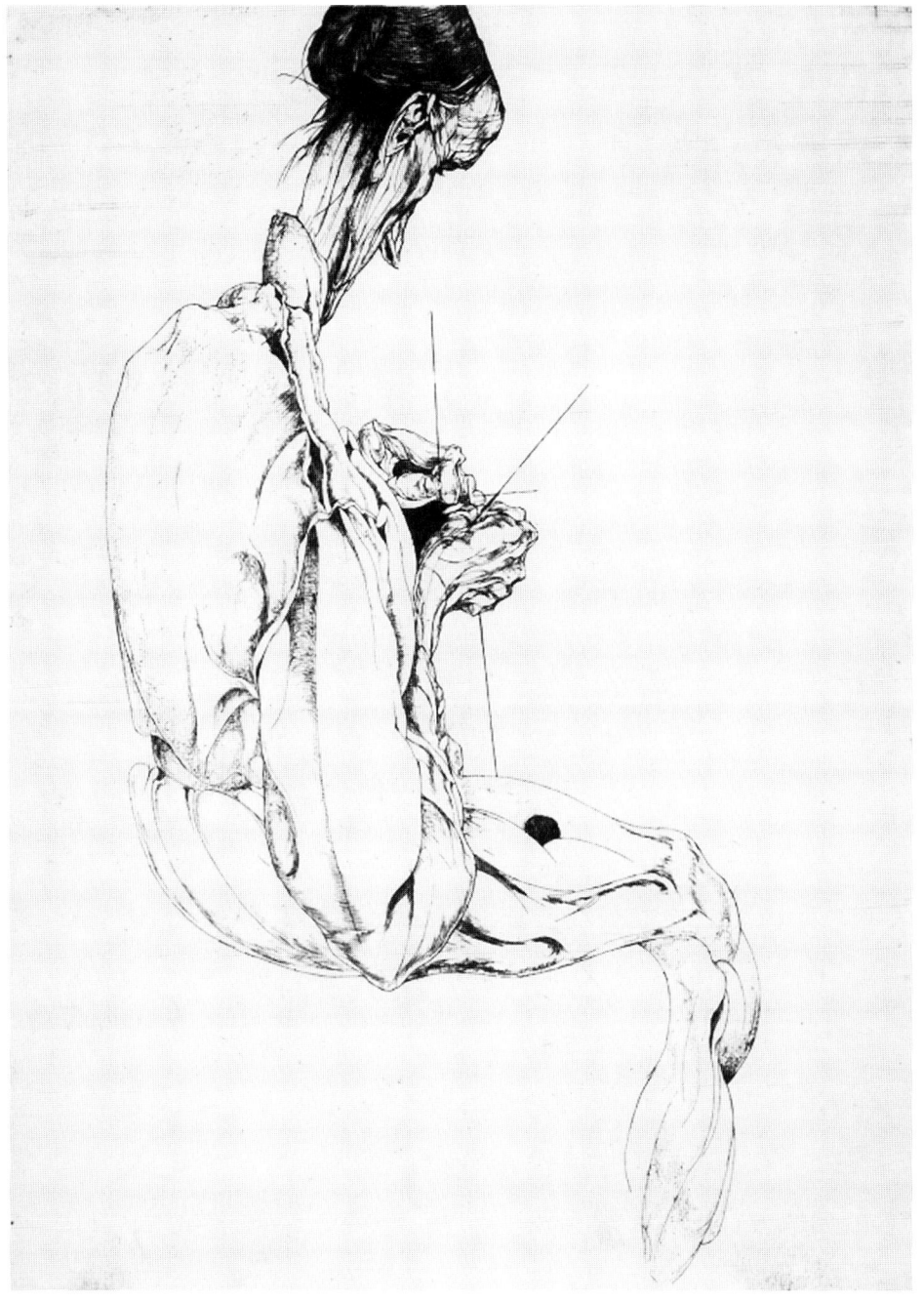

28
Breidster [Knitter], 1925

Etching
400 x 296

Bibliothèque royale Albert 1er,
Cabinet des Estampes, Brussels
Pl. SIII 80311
Le Roy no. 106

29
Tailleur [Tailor], 1925

Etching and drypoint
Inscribed: 1r état 4 épreuves
374 x 320

J. De Bruycker Collection
Le Roy no. 110

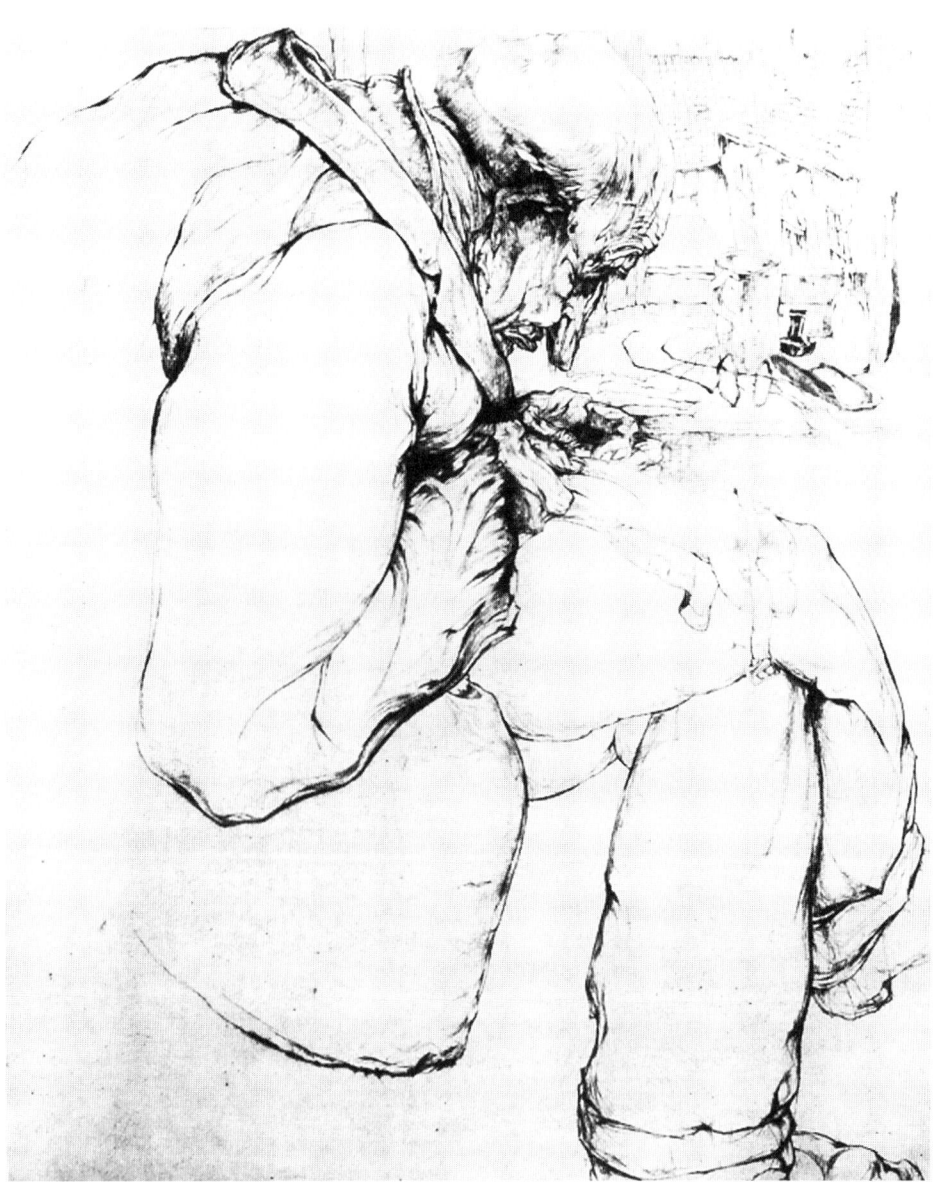

30
Le vieux Bruxelles [Old Brussels], 1925

Etching and aquatint
315 x 206

J. De Bruycker Collection
Le Roy no. 118 (proof)

31

[*L' Echaugette quai St. Pierre, Gand*,
The Warming Tower, Quai St. Pierre,
Ghent], 1925

Etching
Inscribed: 1r état 4 épreuves
400 x 300

J. De Bruycker Collection
Le Roy no. 122

32

L' Echaugette quai St. Pierre, Gand
[The Warming Tower, Quai St. Pierre, Ghent], 1925

Etching and aquatint
Inscribed: Très cordiallement à
Madame E. Bertainchaud 2–12–36
343 x 245

J. De Bruycker Collection
Le Roy no. 122

33

l'Église St.-Nicolas à Gand
[Church of St. Nicholas, Ghent], 1928

Etching and drypoint
Inscribed: existe 2 épreuves d'état
(plaque detruit)
587 x 448

J. De Bruycker Collection
Le Roy no. 151 (variant)

34
Kermesse [Kermiss], 1928

Etching and aquatint
347 x 357

Bibliothèque royale Albert 1er,
Cabinet des Estampes, Brussels
Pl. SIV 14963
Le Roy no. 153

35
Jardin [Garden], 1928

Etching
20 x 16

Bibliothèque royale Albert 1er,
Cabinet des Estampes, Brussels
Fol. SIII 80302
Le Roy no. 156

36
L'Atelier [The Studio], 1928

Etching
Inscribed: épreuve 1r état
317 x 235

J. De Bruycker Collection
Le Roy no. 157

37

De ma fenêtre [From My Window], 1928

Etching
Inscribed: épreuve 1r état
18 x 14

Bibliothèque royale Albert 1er, Cabinet des Estampes, Brussels
Fol. SIII 80335
Le Roy no. 158

38
La Cathédrale d'Anvers
[Antwerp Cathedral], 1929

Etching and aquatint
611 x 488

Museum voor Schone Kunsten, Ghent
1948 T
Le Roy no. 163

39
Rouen, 1930

Etching and aquatint
Inscribed: 1er état
604 x 494

Museum voor Schone Kunsten, Ghent
1932 DH
Le Roy no. 166

40

Bruxelles jour de Bourse [Stock Exchange Day, Brussels], 1930

Etching and aquatint
502 x 405

Bibliothèque royale Albert 1er,
Cabinet des Estampes, Brussels
Max. SIV 14969
Le Roy no. 167

41
[*l'Aquafortist*, The Etcher], 1933

Drypoint
Inscribed: 1r état
216 x 156

J. De Bruycker Collection

42

Sites et visions de Gand 8, Une maison bourgeoise [Sites and Visions of Ghent 8, a Bourgeois House], 1932

Etching
212 x 153

Fogg Art Museum, Harvard University Art Museums: Gift of George Sarton
M10987

43

Sites et visions de Gand 12, Eglise Saint-Michel [Sites and Visions of Ghent 12, The Church of St. Michael], 1932

Etching
212 x 157

Fogg Art Museum, Harvard University Art Museums: Gift of George Sarton
M10991

44

Sites et visions de Gand 18, Sluisken (Quai de l'écluse) [Sites and Visions of Ghent 18, Sluice (Floodgate Quay)], 1932

Etching
237 x 160

Fogg Art Museum, Harvard University Art Museums: Gift of George Sarton
M10997

45

[*Dessin autoportrait devant l'église St.-Nicolas Gand*, Drawn Self-Portrait in front of the Church of St. Nicholas in Ghent], ca. 1936

Charcoal, conté crayon, and graphite
437 x 310

J. De Bruycker Collection

46
Model, 1936

Graphite and conté crayon
576 x 395

J. De Bruycker Collection

47

De Sint-Niklaaskerk te Gent [The Church of St. Nicholas, Ghent], 1937

Graphite and conté crayon
640 x 488

Museum voor Schone Kunsten, Ghent
1972 AQ

48

Jeune couple, Terrasse du Wilson Gand [Young Couple on the Terrace of the Wilson], 1941
From the portfolio *Gens de chez nous* [People from Around Here]

Etching
First state
253 x 222

J. De Bruycker Collection

49

Oorlogsetsen en -schetsen
[Wartime Etchings and Drawings]
no. 1, 1940–45, published 1975

Etching

210 x 183
Spencer Museum of Art: Gift of
A. De Bruycker, daughter of the artist,
and J. De Bruycker
1994.21.1

50
Oorlogsetsen en -schetsen
[Wartime Etchings and Drawings]
no. 3, 1940–45, published 1975

Etching
175 x 257

Spencer Museum of Art: Gift of
A. De Bruycker, daughter of the artist,
and J. De Bruycker
1994.21.3

51
Oorlogsetsen en -schetsen
[Wartime Etchings and Drawings]
no. 5, 1940–45, published 1975

Etching
168 x 288

Spencer Museum of Art: Gift of
A. De Bruycker, daughter of the artist,
and J. De Bruycker
1994.21.5

BIBLIOGRAPHY

Names and titles with particle are alphabetized by the particle (for example, De Bruycker under D and van de Woestijne under V), while titles of publications, groups etc. with articles are alphabetized by the noun (for example, Le Folklore under F).

Anonymous. "Belgische Kunst te Rotterdam." *Nieuwe Rotterdamsche Courant* 74 no. 116 (28 April 1917): B1.

———. "News and Views. Horrors of War Theme of Many De Bruycker Etchings." *Chicago Tribune* (15 October 1922).

———. "Jules de Bruycker chez Giroux à Bruxelles." *Sélection, Chronique de la vie artistique* 2 no. 54–56 (June 1922): 160.

Baedeker, Karl. *Belgium and Holland Including the Grand-Duchy of Luxembourg.* Leipzig: Karl Baedeker, 1910.

Baillieul, Beatrix, and Anne Duhameeuw. *Een Stad in Opbouw. Gent voor 1540.* Tielt: Lannoo, 1989. [see also Dambruyn]

Berg, C. "Bos(s)chère, Jean de." In *Nationaal Biagrafisch Woordenboek* 8, 90–96. Brussels: Paleis der Academiën, 1979.

Bibliothèque royale Albert 1er, Brussels. *Prenten van Jules De Bruycker* exhibition catalogue, 1970.

Bonnel, P.J.J. "Herinneringen aan Juul de Bruycker." *De Autotoerist* 9 no. 18 (16 September 1956): 1051–64.

Bouwen door de eeuwen heen. Inventaris van het cultuurbezit. In *België. Architectuur,* 4na, n.p. Stad Gent: Brepols, 1976.

Boyens, Piet. *Sint-Martens-Latem. Kunstenaarsdorp in Vlaanderen.* Tielt: Lannoo, 1992.

———. (translation of above titile) *Flemish Art, Symbolism to Expressionism at Sint-Martens-Latem.* Tielt: Lannoo, 1992.

Buenger, Barbara C. "Max Beckmann in the First World War." In Rumold and Werckmeister, 237–280.

Buschmann, P. "Belgian Artists in England." *The Studio* 63 no. 261 (15 December 1914): 183–210.

Capiteyn, André. *Gent in Weelde Herboren. Wereldtentoonstelling 1913.* Ghent: Stadsarchief, 1988.

Chabot, Georges. "Arabesken rond het werk van Jules De Bruycker." *Kunst* 4 2/3 (1933): 33–71.

———. "Jules De Bruycker." *Kultureel Jaarboek van de Provincie Oost-Vlaanderen 17.* Ghent: 1963, pp. 96–176.

———. "Jules De Bruycker." *L'Art et la Vie* 2 (15 February 1934): 43–62.

———. "Jules De Bruycker." *Revue des Amateurs* 8 (15 March 1946): 226–31.

———. "Jules de Bruycker. Aquafortiste gantois." *La Revue d'Art* 28 no. 10 (October 1926): 75–86.

———. "Jules De Bruycker etser." *Onze Kunst* 44 nos. 11–12 (November–December 1926): 63–73.

———. "En marge d'une exposition Jules De Bruycker. Un énorme labeur d'esprit, de coeur, et de main." *Le Bien Public* 85 no. 40 (9 February 1938): 1.

Cork, Richard. *A Bitter Truth. Avant-Garde Art and the Great War.* New Haven: Yale University Press, 1994.

Crick, Jef. "De Gentsche Meester-Etser Jules de Bruycker." *De Stad Antwerpen* 4 no. 50 (26 February 1932): 1336–37.

Dambruyn, Johan, Guido Jan Bral, Aletta Rambout, and Dirk Laport. *Een Stad in Opbouw. Gent van 1540 tot de Wereldtentoonstelling van 1913.* Tielt: Lannoo, 1992. [see also Baillieul]

De Bosschère, Jean. *Beasts and Men.* London: William Heinemann and New York: Dodd, Mead & Co., 1918.
———. "Bruegel le Drôle et notre gout en peinture." *L'Occident* 122 (January 1913): 7–14, and 123 (February 1913): 59–70.
———. *Christmas Tales of Flanders.* London: William Heinemann and New York: Dodd, Mead & Co., 1917.
———. ["J."] "Exhibition by Belgian Artists, the Dowdeswell Galeries, Juni 1917." *Onze Kunst* 16 no. 32 (July–December 1917): 131–32.
———. "Exhibition of Belgian Art for the Benefit of the Croydon General Hospital. The Art Gallery, Octob.–Novemb. 1916." *Onze Kunst* 16 no. 31 (January–June 1917): 33–34.
———. *Jérome Bosch.* Brussels: Éditions du Cercle d'art, 1947.
———. "Jules De Bruycker, Teekenaar en Etser." *Onze Kunst* 33 (January–June 1918): 1–17.
———. *Portraits d'amis.* Paris: Editions Sagesse, 1935.
———. *Quinten Metsys.* Brussels: Van Oest & Co., 1907.
———. *La sculpture anversoise aux XVe et XVIe siècles.* Brussels: Van Oest & Co., 1909.

De Bos[s]chère, Jean, Michel Manoll, and Guy Le Clec'h. *Franz Helens.* Lyon: Les Écrivains Réunis Armand Henneuse, 1958.

Decavele, Johan, and Bart Doucet, eds. *De Opera van Gent. Het 'Grand Théâtre' van Roelandt, Philastre en Cambon.* Tielt: Lannoo, 1993.

De Coster, Charles. *La Légende et les aventures héroiques, joyeuses et glorieuses d'Ulenspiegel et de Lamme Goedzak au pays de Flandres et ailleurs.* Antwerp: Les Editions du Dauphin, Brussels: Les Editions Robert Sand, Paris: Les Editions G. Crès Cle, 1922.

De Keyser, P. "De Gentsche Folklore in het werk van Julius De Bruycker." *Oostvlaamsche Zanten* 21 no. 1–2 (January–April 1946): 17–21.

De Saegher, R. "J. De Bruycker." *Petite Revue Illustrée de l'Art / de l'Archéologie en Flandre* 3 no. 17 (15 September 1902): 141–42.

De Seyn, Eugène. *Dictionnaire des Ecrivains Belges*, 2 vols. Bruges: Excelsior, 1930–32.
———. *Dictionnaire Historique et Géographique des Communes Belges.* 2 vols. Brussels: A. Bieleveld, 1933.

De Smet, Frédéric. "Jules De Bruycker." *Gand Artistique* 1 no. 9 (1 September 1922): 126–29.
———. "Jules De Bruycker." *L'Art Belge* 4 no. 3 (31 March 1922): 1–4.
———. "Nos Artistes. Mme Cécile Cauterman." *Gand Artistique* 1 no. 1 (1 January 1922): 5–7.

Devoghelaere, Hubert. "Jules De Bruycker." *Kunst* 4 no. 2/3 (1933): 65–71.

Dictionnaire biographique illustré des artistes en Belgique depuis 1830. Brussels: Arto, 1987.

Dutry, Albert. "Jules De Bruycker." *Durendal. Revue Catholique d'Art et de Litérature* 13 (1906): 256–59.

Eeckhout, Paul. "Bruycker, Jules-François De." In *Biographie Nationale* suppl. 16 (vol 44): 130–140. Brussels: Emile Bruylant, 1985–86.
———. *Jules De Bruycker. Peintures, Aquarelles, Dessins*, exhibition catalogue. Ghent: Museum voor Schone Kunsten, 1970.
———. *Retrospectieve Tentoonstelling Albert Baertsoen, 1866–1922*, exhibition catalogue. Ghent: Museum voor Schone Kunsten, 1973.

Encyclopedie van de Vlaamse Beweging. 2 vols. Tielt: Lannoo, 1973–1975.

Estang, Luc, et. al. *Jean De Bosschère l'admirable*. Paris: Au Parchemin d'Antan, 1952.

Evans, Martin Marix. *Ypres in War and Peace*. Andover, Hants: Pitkin, 1992.

Fletcher, John Gould. "The Drawings of Jean De Bosschère." *The Studio* 78 no. 323 (Feb. 1920): 193–201.

Le Folklore dans l'oeuvre de Charles De Coster, special edition of *Le Folklore Brabançon* 7 nos. 37–38 (August–October 1927).

Friedman, Donald Flanell. "Belgian Symbolism and a Poetics of Place." In Goddard, et al., *Les XX and the Belgian Avant-Garde*, 126–38.
———. *The Symbolist Dead City. A Landscape of Poesis*. New York and London: Garland, 1990.

Fris, Victor. *Histoire de Gand depuis les origines jusqu'en 1913*. 2nd edition. Ghent: Gaston de Tavernier, 1930.

Frommer, Arthur. *A Masterpiece Called Belgium*. Brussels: Sabena, 1989.

Goddard, Stephen, et al. *Les XX and the Belgian Avant-Garde: Prints, Drawings, and Books, ca. 1890*. Lawrence: Spencer Museum of Art, 1992.

Grote Winkler Prins Encyclopedie. 25 vols. Amsterdam and Brussels: Elsevier, 1979–89.

Guislain, Albert. "La balade gantoise." *L'Indépendence Belge* 108 no. 359 (25 December 1938): 1,5.
———. "Gand et Jules De Bruycker." *L'Indépendence Belge* 108 no. 361 (27 December 1938): 1, 3.
———. "Jules De Bruycker aquafortiste et poète." *L'Indépendence Belge* 109 no. 3 (3 January 1939): 1, 3.
———. *Le Palais de Justice ou les confidences du mammouth*. Brussels: Editions du Cheval de Bois, 1935.

H.L. "Sites et Visions de Gand." *Le Soir Illustré* 5 no. 220 (7 May 1932): 9.

Hardie, Martin. *Frederick Goulding, Master Printer of Copper Plates*. Stirling: Eneas Mackay, 1910.

Hasselt, Provinciaal Begijnhof. *Jules De Bruycker*, exhibition catalogue n.d.

Heino, Hannema-De Steurs Fundatie. *De Collectie Schilderijen, Beeldhouwwerken, Tekeningen en Grafiek De Graaff-Bachiene*, exhibition catalogue, 1972.

Hellens, Franz. *Documents secrets (1905–1956). Histoire sentimentale de mes livres et de quelques amitiés*. Paris: Éditions Albin Michel, 1958.
———. *En Ville Morte. Les Scorcies*. Brussels: G. Van Oest & Co., 1906.
———. *Les Hors-le-vent*. Brussels: Librairie moderne, 1912.
———. "Jules De Bruycker." *L'Art Moderne* 27 no. 1 (6 January 1907): 1–2.
———. "La Puissance de Gand," *L'Art vivant* 6 no. 136 (15 August 1930): 650–59.

Hoozee, Robert. *Veertig Kunstenaars Rond Karel van de Woestijne*, exhibiton catalogue. Ghent: Museum voor Schone Kunsten, 1979.

Hoozee, Robert, Sabine Bown-Taevernier and J.F. Heijbroeck. *James Ensor, Dessins et Estampes*. Antwerp: Fonds Mercator, 1987.

Hymans, Henri. "Bruycker, Jules De." In Thieme and Becker, *Allgemeines Lexikon der Bildende Künstler*, vol. V: 153. Liepzig: Verlag von E.A. Seemann, 1911.

Imperial War Museum, London. *A Concise Catalogue of Paintings, Drawings, and Sculpture of the First World War 1914–1918*, 2nd edition, 1963.

Khnopff, Fernand [K, F]. "Studio Talk / Brussels." *The Studio* 62 no. 254 (June 1914): 72–75.

Klinkenberg, Jean-Marie. *Charles De Coster*. Brussels: Éditions Labor (Une Livre Une Oeuvre 2), 1985.

Kluyskens, Pierre. *De Gentenaar, Katoliek Dagblad* 94 no. 188 (28 April 1969): 3.
———. "Jules De Bruycker 1870–1945." *De Gentenaar, Katoliek Dagblad* 81 no. 14 (15 January 1956): 3.

Langui, Emiel. "Afscheid van Jules De Bruycker. † 6 September 1945." *Zondagspost, onafahankelijk weekblad voor politiek en cultuur* 1 no. 38 (23 September 1945): 8.

Langui, Emiel, and Achilles Mussche. "Jules De Bruycker Zeventig Jaar. Gent 29 Maart 1870." *Vooruit* 56 no. 89 (31 March 1940): 8.

Laran, Jean. *L'Estampe*, 2 vols. Paris: Presses Universitaires, 1959.

Lebeer, Louis. *Etskunst*. Brussels: Standaard and Amsterdam: van Kampen & Zoon, 1929.
———. *De Graphischen Kunsten in Vlaanderen*, 376–379. Amsterdam-Antwerp, 1932.

Le Roy, Grégoire. *L'Oeuvre gravé de Jules De Bruycker*. Brussels: Nouvelle Société Éditions, 1933.

Marechal, Dominique. *Collectie Frank Brangwyn*. Bruges: Stedelijke Musea and Generale Bank, 1987.

Marijnissen, Roger. "Artiest chargeerde te veel of te weining. De Bruycker is het best als hij het minst De Bruycker is." *De Standaard* (30 April 1970).

Marlow, Georges. "Chronique de Belgique." *Mercure de France* 45 no. 572 (15 April 1922): 519–20.

McC., L.M. "De Bruycker, etcher of Ghent, Introduced to America." *Christian Science Monitor* (20 October 1922).

Murez, Jos. "Jules De Bruycker 1870–1945." *Vooruit* 86 no. 276 (26 March 1970): 6.
———. "Jules De Bruyckers 'oude verkoopster'." *Vooruit* 88 no. 22 (27 May 1971): 7.

Museum voor Schone Kunsten, Ghent. *Duizend Jaar Kunst en Cultuur*, exhibition catalogue, 3 vols., 1975.

Mussche, Achilles. *Gent en zijn etser-teekenaar Jules De Bruycker*. Oude-God-Antwerp: Boekengilde Die Poorte, 1935–36.

Polak, Betinna. *Het Fin-de-siècle in de Nederlandse Schilderkunst*. 's-Gravenhage: Martinus Nijhoff, 1955.

Pudles, Lynne. "Fernand Khnopff, Georges Rodenbach, and Bruges, the Dead City." *Art Bulletin* 74 no. 4 (December 1992): 637–654.

Putnam, Samuel. *The World of Jean de Bosschère*. London: The Fortune Press, 1932.

Roelandt, Oscar [R]. "Jules De Bruycker 1870–29 — Maart–1930." *Vooruit* 44 no. 88 (30 March 1930): 3.

Rumold, Rainer and O.K. Werckmeister, eds. *The Ideological Crisis of Expressionism: The Literary and Artistic German War Colony in Belgium 1914–1918*. Columbia, S.C.: Camden House, 1990.

Schepens, Luc. *Kroniek van Stijn Streuvels 1871–1969*. Bruges: Orion, 1971.

Shapiro, Theda. *Painters and Politics. The European Avant-Garde and Society, 1900–1925*. New York, Oxford, Amsterdam: Elsevier, 1976.

Simons, Ludo. "Tijl Uilenspiegel." In *Encyclopedie van de Vlaamse Beweging*, vol 2: 1665–66.

Simons, Walter, Guido Jan Bral, Jan Caudron, and Johan Bockstaele. *Het Pand. Acht eeuwen geschiedenis van het oud Dominicanenklooster te Gent*. Tielt: Lannoo, 1991.

Stedelijk Museum Sint-Niklaas. *Jules De Bruycker*, exhibition catalogue, 1993.

Somers, Marc., et. al. *Karel Van de Woestijne, 1878–1929*, exhibition catalogue. Brussels: Koninklijke Bibliotheek Albert 1er, 1979.

Streuvels, Stijn. *In oorlogstijd. Het volledige dagboek van de eerste wereldoorlog*. Bruges: Orion and Nijmegen: B. Gottmer, 1979.

van de Woestyne, Gustave. *Karel en ik, herinneringen*. Brussels: Elsevier Manteau, 1979.

van de Woestijne, Karel. "Jules De Bruycker." *Elseviers geillustreerd Maandschrift* (April 1912), as published in *Versameld Werk* (vol 4) *Beschouwen over Literatuur en Kunst*, 893–936.
———. "Jules De Bruycker." I: *Nieuwe Rotterdamse Courant* 79 no. 56 (26 February 1922) Ochtenblad A, 1–2; II: 79 no. 57 (27 February 1922) Avondblad B, 1–2; III: 79 no. 58 (28 February 1922) Avondblad B, 1.
———. *Versameld Werk*, 8 vols. Brussels: Manteau, 1947–1950.
———. "De Vlaamsche Primitieven, hoe ze waren to Brugge [1902]." In *verzameld werk* IV, 9–130.

van de Woestijne, Karel, and Herman Teirlinck. *De Leemen Torens*, 2 vols. Rotterdam: Nijgh & Van Ditmar's, 1928.

Van Lerberge, Raf. *De Geschiedenis van Bond Moyson. De betekenis van de mutualiteit in de ontwikkeling van de Gentse arbeidersbeweging*. Ghent: Archief en Museum van de socialistische Arbeidersbeweging, 1979.

Vanzype, Gustave. "Notice sur Jules De Bruycker, Membre de l'Académie." *Annuaire de l'académie royale de Belgique / Jaarboek van de Koninklijke Belgische Academie* 116, 1950: 167–85.

Walker, R.A. "Jules De Bruycker." *The Print Collector's Quarterly*, London, 1934, I: 37–58.

Warmoes, Jean. "Boschère (Jean de)." *Biographie nationale* supplément vol. 14 (1981): 56–82.
———. *Jean De Bosschère. Le centenaire de sa naissance*. Brussels: Bibliothèque royale Albert 1er, 1978.

Wijngaert, Frank van den. *Jules De Bruycker*. Antwerp: De Sikkel, 1948.

Willis, Alfred. *Flemish Renaissance Revival in Belgian Architecture (1830–1930)*. Ph.D. Thesis, Columbia University, 1984.

Winternitz, Emanuel. "Bagpipes and Hurdy-Gurdies in their Social Setting." *Metropolitan*

Museum of Art Bulletin n.s. 11, no. 1 (Summer, 1943): 56–83.

Wurfgain, M.L. *Belgische grafici: Rops, Ensor, De Bruycker, Smits*, exhibition catalogue. Rotterdam: Museum Boymans Van Beuningen, 1967.

Young, Andrew McLaren, Margaret MacDonald, and Robin Spencer. *The Paintings of James McNeill Whistler*, 2 vols. New Haven: Yale University Press, 1980.

Zerck, Arch. J. "De Reanimatie van de Gentse Stadswijk het 'Patershol'." *Kultureel Jaarboek voor de Provincie Oost-Vlaanderen* vol. II: *Het Pand van de Geschoeide karmelieten te Gent en de Reanimatie van de Wijk "het Patershol,"* 1973.

INDEX

Names and titles with particles are alphabetized by the particle (for example, De Bruycker under D and van de Woestijne under V) while titles of publications, groups, etc. with articles are alphabetized by the noun (for example, Le Folklore under F)

Academy of Fine Art, Ghent, see *Ghent*
Alba, Duke of, 24, 28
Alijn, Jacobus, 31
Amien, 33
Anspach, Jules, 6
Antwerp, 4, 27, 31, 33
Apollinaire, Guillaume, 22
Artaud, Antonin, 25
Baarle, 24
Baertsoen, Albert, 14, 20, 34
Baldwin Iron Arm, Count of Flanders, 3
Battle of the Golden Spurs, 3
Beardsley, Aubrey, 25
Beguines, 3
Berlin, 37
Binding, Rudolph, 28
Black Prince of Wales, 3
Boccioni, Umberto, 21
Bonnel, Peter, 4, 27
Borluut, Jan, 3
Bosch, Hieronymus, 23, 26, 32
Bourges, 33
Brangwyn, Frank, 20, 22–23, 29
Braque, Georges, 22
Braun, Emile, 6
Broederlijke Maatschappij van Gentsche Wevers, 3
Brothers of the Common Life, 24
Bruegel, Pieter the Elder, 1, 23, 24, 26, 27, 28, 37
Bruges, 3, 17, 24
Brussels
    Bourse, 32, 33
    Galerie Giroux, 31
    Grafisch Salon, 32
    Maison Dietrich, 14
    Marolles, 19
    Palais de Justice, 19–20, 32
    Ste.-Gudule (cathedral), 32
Canell, Théo, 8

Chabot, Georges, 13, 20, 21, 31, 35, 36
Chemical warfare, 27
Chicago Art Institute, 31
Christus, Petrus, 24
Claus, Emile, 21
Cotett, Charles, 25
Dadaists, 21
Daeye, Hippolyte, 21
Daumier, Honoré, 13
de Bosschère, Jean, 4, 25, 26, 27, 32–33
De Bruycker, Jules—titles of works by
    l'Aquafortist, 34; cat. 41
    L'Atelier, 34; cat. 36
    Breidster, 31; cat. 28
    Bruxelles jour de Bourse, 33; cat. 40
    La Cathédrale d'Anvers, 33; cat. 38
    Confrère, 13, 15; cat. 1
    De ma fenêtre, 34; cat. 37
    Dessin autoportrait devant l'église St. Nicolas Gand, 34; cat. 45
    L'Echaugette quai St. Pierre, Gand, 33; cat. 31, cat. 32
    l'Église St. Nicolas à Gand, 34–35; cat. 33
    En Ville Morte, 7; cat. 6, cat. 7
    Étude, 18
    Étude-marché, 17
    Frans Masereel, 15
    Gens de chez nous, 36; see also *Jeune couple*
    Gens pas de chez nous, 37
    Jacobus Alijn or *De bedelaar*, 31; cat. 27
    Jacobus Alijn (le poilu), 31; cat. 25
    Jardin, 34; cat. 35
    Jeune couple, Terrasse du Wilson Gand from *Gens de chez nous*, 35–36; cat. 48
    Jour de Marché à Gand, 17; cat. 3
    Kermesse, 32, 34; cat. 34
    Kultur!, 26, 27; cat. 20
    La Légende […]d'Ulenspiegel…, 23–24, 28–29, 31–32; cat. 26

La Maison Jean Palfijn Gand, 18–19, 20; cat. 15
De man van l'belfort, 20; cat. 17
Marchand de bric-à-brac, 17; cat. 12
Mendiants, see *Kermesse*
Model, 34; cat. 46
La Moisson, 27; cat. 21
La montage du Dragon sur le beffroi de Gand (Belgique), 20; cat. 18
Oorlogs etsen en -schetsen, 37; cat. 49, cat. 50, cat. 51
Le paradis (théâtre), 17; cat. 8
Patershol artiste, 30–31
Petite Ville Nerveuse, see *Tweeslachtige Stad*
Piccadilly Londres, 21; cat. 23
La Porte St. Denis, Paris, 33
De Rolweg te Brugge, 17; cat. 14
Rond het s'Graven Kasteel te Gent, 19, 20; cat. 16
Rouen, 33; cat. 39
Ruelle (Gand), 15; cat. 2
Ruelle (Patershol), 15; cat. 4
De Sint-Niklaaskerk te Gent, 35; cat. 47
Sites et visions de Gand, 25, 34; cat. 42, cat. 43, cat. 44
De Slechte Maere, 27, 28, 29, 32; cat. 24
Tailleur, 31; cat. 29
Théâtre (Lohengrin), 17; cat. 11
La Tranchée, 27; cat. 22
Tweeslachtige Stad, 34
L'Uilenkot, 17; cat. 9
Veergrepe, 15; cat. 5
Le vieux Bruxelles, 32; cat. 30
Vieux marché à Gand, 17; cat. 10
Vieux marché en Flandre, 17; cat. 13
Weer klepte de Dood over Vlaanderenland, 23, 26, 28; cat. 19
De Coster, Charles, 24, 28, 29, 31–32, 34
de Graaff-Bachiene, Jacob and Louise, 4
de Kalle, Cies (pseudonym of Georges vande Walle), 13, 15, 31
de La Fresnaye, Roger, 21
De Leyn, Raphaëlle, 21
Delstanche, Albert, 21
Delvin, Jean-Joseph, 8, 14
de Praetere, Jules, 9, 10, 13, 17
de Sadeleer, Valerius, 17, 21, 24
Duchamp-Villon, Raymond, 21
Dufy, Raoul, 22
Elsene, 13
Elskamp, Max, 25
Ensor, James, 25, 33
Fabry, Emile, 21

La Feuille, 21
Flemish Movement, 1, 4
Flemish Primitives, 10, 24
Ghent, 3–6
    Academy of Fine Art, 7, 8, 14
    Ballenstraat, 15
    Beguinage, 3
    Bloedsteeg, 6
    Café Wilson, 35
    Cloth Hall, 3, 6
    Gravenkasteel, see *Gravensteen*
    Gravensteen [Castle of the Counts], 3, 6, 9, 11, 13, 15, 17
    Gruisberg, 6
    Haringsteeg, 15
    Jan Breydelstraat, 7
    Kalletje, 6
    Kattenberg, 6
    Kraanlei, 18
    Luizengevecht, 6
    Nieuwpoortje, 6
    Pand, Carmelite, (patershol), 11
    Pand (Dominican), 31
    Patershol, 8, 9–14, 15, 24, 31
    Reep, 6
    Rolleken, 11
    Serpentstraat, 6
    Sint-Baafs (abbey), 3
    Sint-Baafs (cathedral), 3, 6
    Sint-Jacobs, 3, 17
    Sint-Jans, 3
    Sint-Michiels, 3, 11, 34
    Sint-Niklaas, 3, 6, 24, 34–35
    Sint-Pieters, 3
    Sint-Pieterskaai, 33
    Sint-Veerleplein, 17
    Veer, 6
    Vooruit, 4
    Vridagmarkt, 3
    Vrouwebroers, 11
Golding & Co., see *Goulding Ltd.*
Goulding Ltd., 23, 29
Gounod, Charles François, 11, 12
Goya, Francisco, 19
Groenendael, 24
les Gueux, 28
Guislain, Albert, 19
Haussmann, Baron George, 6
Hellens, Franz, 4, 5, 7, 13, 32
Hoozee, Robert, 9
Hundred Years War, 3
Ieper, 3, 27–28
Images d'Épinal, 22
Ixelles, see *Elsene*

Khnopff, Fernand, 24
Kirchner, Ernst Ludwig, 22
Kloos, Willem, 25
Kokoschka, Oskar, 22
Koninklijke Vlaamse Academie voor Taal- en Letterkunde, 4
Kortrijk, 3
Kropotkin, Prince Pyotr Alekseyevich, 10, 13
Lambotte, Mr., 23
Langemarck, 28
"Language problem," 4, 34
Leie River, 3
le Roy, Grégoire, 4, 5, 34
Lieve River, 3, 6
Lier, 27
Lippens, Hippolyte, 6
London, 20, 21, 31
Louis, Count of Flanders, 3
Lubok prints, 22
Maatschappij der Noodlijdende Broeders, 4
Macke, Auguste, 21
Maeterlinck, Maurice, 5, 24
Maison Dietrich, see under *Brussels*
Malevich, Kasimir, 22
Marc, Franz, 21
Marijnissen, Roger, 32
Marolles, see under *Brussels*
Masereel, Frans, 5, 15, 21, 33
Memling, Hans, 24
Mercure de France, 13
Mertens, Charles, 21
Meryon, Charles, 33
Metsys, Quinten, 24, 25
Minne, Georges, 5, 21, 24, 25
Minne, Mrs. Georges, 24
Moyson, Emiel, 4
Mussche, Achilles, 4, 5, 17
De Nieuwe Gids, 25
Nieuwe Rotterdamsche Courant, 23, 24
L'occident, 25
Opsomer, Isidore, 21
Les Primitifs flamands, see *Flemish Primitives*
Oudenaard, 10
Palais de Justice, see under *Brussels*
Palfijn, Jan, 18
Paris, 33
Pater, Walter, 25
Patershol, see under *Ghent*
Piranesi, Giovanni Battista, 33

Puvis de Chavannes, Pierre, 11
Rembrandt, 34
Rodenbach, Georges, 7
Rouen, 33
Royal Academy of Belgium, 31, 33
Rousseau, Jean-Jacques, 11
Sant'Elia, Antonio, 21
Saurès, André, 25
Scheldt River, 3
Shakespeare, William, 10
Sint-Martens-Latem, 5, 10, 15, 24, 25
Schiller, Johann Christoph Friedrich von, 11, 12
socialism, 1, 5
Stevens, Gustave-Max, 21
Streuvels, Stijn, 9, 10, 13, 24
Surrealists, 25
tapissier, 7, 8, 13, 31
Teirlinck, Herman, 9, 13
Les Temps nouveaux, 12, 13
Tijtgadt, Louis, 8, 14
Ulenspiegel, Tijl, 23–24, 28, 29, 31–32
Van Artevelde, Jacob, 3
Van Campenhout, 14
van der Goes, Hugo, 4, 25
Van der Swalm, 13
van der Weyden, Roger, 24
vande Walle, Georges (birthname of Cies de Kalle), 13
van de Woestijne, Gustav, 5, 10, 21, 24, 25
van de Woestijne, Karel, 4, 5, 8, 9, 10, 11–13, 17, 18, 21, 24, 37
van Eyck, Hubert and Jan, 3, 4, 24
van Herrewege, René, 8, 20
van Loo, Fritz, 14
Van Nu en Straks, 10
van Ruusbroec, Jan, 24
van Rysselberghe, Théo, 5
van Vooren, Emile, 13
Venice Biënnale, 20
Vooruit, 4
weaving industry, 3
Whistler, James Abbott McNeill, 21, 29, 34
World Exposition, 1913, 6
World War I, 8, 21–29, 37
World War II, 36–37
Ypres, see *Ieper*
Zille, Heinrich, 17